Studies in Fuzziness and Soft Computing

Editor-in-chief

Prof. Janusz Kacprzyk
Systems Research Institute
Polish Academy of Sciences
ul. Newelska 6
01-447 Warsaw, Poland
E-mail: kacprzyk@ibspan.waw.pl
http://www.springer.de/cgi-bin/search_book.pl?series=2941

James J. Buckley

Fuzzy Probabilities

New Approach and Applications

With 36 Figures
and 17 Tables

Physica-Verlag

A Springer-Verlag Company

Professor James J. Buckley
University of Alabama at Birmingham
Mathematics Department
Birmingham, AL 35294
USA
buckley@math.uab.edu

ISBN 978-3-642-86788-0   ISBN 978-3-642-86786-6 (eBook)
DOI 10.1007/978-3-642-86786-6
Library of Congress Cataloging-in-Publication Data applied for
Die Deutsche Bibliothek – CIP-Einheitsaufnahme
Buckley, James J.: Fuzzy probabilities: new approach and applications; with 17 tables / James J. Buckley. – Heidelberg; New York: Physica-Verl., 2003
   (Studies in fuzziness and soft computing; Vol. 115)

Physica-Verlag Heidelberg New York
a member of BertelsmannSpringer Science+Business Media GmbH

© Physica-Verlag Heidelberg 2003

SPIN 10893772   88/2202-5 4 3 2 1 0 – Printed on acid-free paper

To Julianne and Helen.

Contents

Chapter 1

Introduction

1.1 Introduction

The first think to do is to explain what is our "new approach" and how it fits into the area of uncertain probabilities. We first consider a very simple example using interval probabilities. Let $X = \{x_1, x_2, x_3\}$ be a finite set and let P be a probability function defined on all subsets of X with $P(\{x_i\}) = a_i$, $1 \leq i \leq 3$, $0 < a_i < 1$ all i and $\sum_{i=1}^{3} a_i = 1$. X together with P is a discrete (finite) probability distribution. In practice all the a_i values must be known exactly. Many times these values are estimated, or they are provided by experts. We now assume that some of these a_i values are uncertain and we will model this uncertainty using intervals. Suppose we estimate a_1 as 0.2 ± 0.1, $a_2 = 0.5 \pm 0.2$ and $a_3 = 0.3 \pm 0.1$. Then we would have these probabilities in intervals $a_1 \in [0.1, 0.3]$, $a_2 \in [0.3, 0.7]$ and $a_3 \in [0.2, 0.4]$. What if we now want the probability of the event $A = \{x_1, x_2\}$, it would also be an interval, say $[A_1, A_2]$, and we would compute it as follows

$$[A_1, A_2] = \{a_1 + a_2 | a_1 \in [0.1, 0.3], a_2 \in [0.3, 0.7], a_3 \in [0.2, 0.4], a_1 + a_2 + a_3 = 1\}. \tag{1.1}$$

We easily see that $[A_1, A_2] = [0.6, 0.8]$ which is not the sum of the two intervals $[0.1, 0.3] + [0.3, 0.7] = [0.4, 1.0]$. We did not get $[0.4, 1.0]$ because of the constraint that the probabilities must add to one. There was uncertainty in the values of the probabilities but there is no uncertainty that there is a probability distribution over X. In this book we will always have the constraint that the probabilities must add to one even though all individual probabilities are uncertain. This is our new approach to fuzzy probability. The above example was for interval probabilities but is easily extended to fuzzy probability.

A fuzzy probability is a fuzzy number composed of a nested collection of intervals constructed by taking alpha-cuts (to be discussed in Chapter 2). So, in our new approach to fuzzy probability the interval $[A_1, A_2]$ will be just one of the alpha-cuts of the fuzzy probability of event A which will be a fuzzy number (also discussed in Chapter 2). When all the probabilities are fuzzy we will still insist that the sum of all the individual probabilities is one. This will produce what we call "restricted fuzzy arithmetic".

The other part of our approach to fuzzy probability theory is our method of dealing with fuzzy random variables. A fuzzy random variable is just a random variable with a fuzzy probability mass function (discrete case), or a fuzzy probability density function (the continuous case). Consider a random variable R_1 with a binomial probability mass function $b(n,p)$ and another random variable R_2 with a normal probability density function $N(\mu, \sigma^2)$. R_1 is a discrete fuzzy random variable when p in $b(n,p)$ is fuzzy and R_2 is a continuous fuzzy random variable when μ and/or σ^2 are fuzzy. These parameters usually must be estimated from some random sample and instead of using a point estimate in $b(n,p)$ for p, or a point estimate for μ and σ^2 in the normal density, we propose using fuzzy numbers constructed from a set of confidence intervals. This procedure of employing a collection of confidence intervals to obtain a fuzzy estimator for a parameter in a probability distribution is discussed in more detail in Chapter 2.

The method of finding fuzzy probabilities usually involves finding the maximum, and minimum, of a linear or non-linear function, subject to linear constraints. Our method for accomplishing this will also be discussed in Chapter 2.

However, our method of restricted fuzzy arithmetic is not new. It was first proposed in ([9]–[11]). In these papers restricted fuzzy arithmetic due to probabilistic constraints is mentioned but was not developed to the extent that it will be in this book. Also, in [17] the authors extend the results in [16] to fuzzy numbers for probabilities under restricted fuzzy arithmetic due to probabilistic constraints similar to what we use in this book. But in [17] they concentrate only on Bayes' formula for updating prior fuzzy probabilities to posterior fuzzy probabilities.

This paper falls in the intersection of the areas of imprecise probabilities ([12],[13],[18], [20]-[23]), interval valued probabilities ([7],[16],[24]) and fuzzy probabilities ([5],[6],[8],[19],[25],[26]). Different from those papers on imprecise probabilities, which employ second order probabilities, possibilities, upper/lower probabilities, etc., we are using fuzzy numbers to model uncertainty in some of the probabilities, but we are not employing standard fuzzy arithmetic to combine the uncertainties. We could use crisp intervals to express the uncertainties but we would not be using standard interval arithmetic ([14],[15]) to combine the uncertainties. We do substitute fuzzy numbers for uncertain probabilities but we are not using fuzzy probability theory to propagate the uncertainty through the model. Our method is to use fuzzy numbers for imprecise probabilities and then through restricted fuzzy algebra calculate other fuzzy probabilities, expected values, variances, etc.

It is difficult, in a book with a lot of mathematics, to achieve a uniform notation without having to introduce many new specialized symbols. Our basic notation is presented in Chapter 2. What we have done is to have a uniform notation within each section. What this means is that we may use the letters "a" and "b" to represent a closed interval $[a, b]$ in one section but they could stand for parameters in a probability distribution in another section.

We will have the following uniform notation throughout the book:

(1) we place a "bar" over a letter to denote a fuzzy set ($\overline{A}$, $\overline{B}$, etc.);

(2) an alpha-cut is always denoted by "α";

(3) fuzzy functions are denoted as $\overline{F}$, $\overline{G}$, etc.;

(4) **R** denotes the set of real numbers; and

(5) P stands for a crisp probability and $\overline{P}$ will denote a fuzzy probability.

The term "crisp" means not fuzzy. A crisp set is a regular set and a crisp number is a real number. There is a potential problem with the symbol "$\leq$". It usually means "fuzzy subset" as $\overline{A} \leq \overline{B}$ stands for $\overline{A}$ is a fuzzy subset of $\overline{B}$ (defined in Chapter 2). However, also in Chapter 2 $\overline{A} \leq \overline{B}$ means that fuzzy set $\overline{A}$ is less than or equal to fuzzy set $\overline{B}$. The meaning of the symbol "$\leq$" should be clear from its use, but we shall point out when it will mean $\overline{A}$ is less that or equal to $\overline{B}$. There will be another definition of "$\leq$" between fuzzy numbers to be used only in Chapter 14.

Prerequisites are a basic knowledge of crisp probability theory. There are numerous text books on this subject so there no need to give references for probability theory.

No previous knowledge of fuzzy sets is needed because in Chapter 2 we survey the basic ideas needed for the rest of the book. Also, in Chapter 2 we have added the following topics: (1) our method of handling the problem of maximizing, or minimizing, a fuzzy set; (2) how we propose to order a finite set of fuzzy numbers from smallest to largest; (3) how we find fuzzy numbers for uncertain probabilities using random samples or expert opinion; (4) how we will use a collection of confidence intervals to get a fuzzy number estimator for a parameter in a probability distribution; (5) how we will be computing fuzzy probabilities; and (6) our methods of getting graphs of fuzzy probabilities.

Elementary fuzzy probability theory comprises Chapter 3. In this chapter we derive the basic properties of our fuzzy probability, the same for fuzzy conditional probability, present two concepts of fuzzy independence, discuss a fuzzy Bayes' formula and five applications. Discrete fuzzy random variables are the topic of Chapter 4 where we concentrate on the fuzzy binomial and the fuzzy Poisson, and then discuss three applications. Applications of discrete fuzzy probability to queuing theory, Markov chains and decision theory follows in Chapter 5,6 and 7, respectively.

Chapter 8 starts our development of continuous fuzzy random variables and we concentrate on the fuzzy uniform, the fuzzy normal and the fuzzy negative exponential. Some applications are in Chapter 8 and an application of the fuzzy normal to inventory control is in the following Chapter 9. We then generalize to joint continuous fuzzy probability distributions in Chapter 10. In Chapter 10 we look at fuzzy marginals, fuzzy conditionals, the fuzzy bivariate normal and fuzzy correlation. Applications of joint fuzzy distributions are in Chapter 11. The first application is for a joint discrete fuzzy probability distribution and the second application is also for a joint discrete fuzzy distribution but to reliability theory. Chapters 12,13 and 15 deal with functions of fuzzy random variables. A law of large numbers is presented in Chapter 14. We finish in Chapter 16 with a brief summary of Chapters 3-15, suggestions for future research and our conclusions.

This book is based on, but considerably expands upon, references [1]-[4]. New material includes the fuzzy Poisson, fuzzy conditional probability, fuzzy independence, many examples (applications) within the chapters, Chapter 7, some of Chapters 9-11, and Chapters 12-15. We briefly discuss the new (unpublished) material in each chapter, whenever the chapter contains published results, at the beginning of each chapter.

1.2   References

1. J.J.Buckley and E.Eslami: Uncertain Probabilities I: The Discrete Case, Soft Computing. To appear.

2. J.J.Buckley and E.Eslami: Uncertain Probabilities II: The Continuous Case, under review.

3. J.J.Buckley and E.Eslami: Fuzzy Markov Chains: Uncertain Probabilities, Mathware and Soft Computing. To appear.

4. J.J.Buckley: Uncertain Probabilities III: The Continuous Case, under review.

5. J.Chiang and J.S.Yao : Fuzzy Probability Over Fuzzy σ-Field with Fuzzy Topological Space, Fuzzy Sets and Systems, 116(2000), pp. 201-223.

6. J.Dunyak, I.W.Saad and D.Wunsch: A Theory of Independent Fuzzy Probability for System Reliability, IEEE Trans. Fuzzy Systems, 7(1999), pp. 286-294.

7. J.W.Hall, D.I.Blockley and J.P.Davis: Uncertain Inference Using Interval Probability Theory, Int. J. Approx. Reasoning, 19(1998), pp. 247-264.

8. C.Huang, C.Moraga and X.Yuan: Calculation vs. Subjective Assessment with Respect to Fuzzy Probability, In: B.Reusch (ed.), Fuzzy Days 2001, Lecture Notes in Computer Science 2206, Springer, 2001, pp. 392-411.

9. G.J.Klir: Fuzzy Arithmetic with Requisite Constraints, Fuzzy Sets and Systems, 91(1997), pp. 147-161.

10. G.J.Klir and J.A.Cooper: On Constrainted Fuzzy Arithmetic, Proc. 5th Int. IEEE Conf. on Fuzzy Systems, New Orleans, (1996), pp. 1285-1290.

11. G.J.Klir and Y.Pan: Constrained Fuzzy Arithmetic: Basic Questions and Some Answers, Soft Computing, 2(1998), pp. 100-108.

12. J.Lawry: A Methodology for Computing with Words, Int. J. Approx. Reasoning, 28(2001), pp. 51-89.

13. L.Lukasiewicz: Local Probabilistic Deduction from Taxonomic and Probabilistic Knowledge-Bases Over Conjunctive Events, Int. J. Approx. Reasoning, 21(1999), pp. 23-61.

14. R.E.Moore: Methods and Applications of Interval Arithmetic, SIAM Studies in Applied Mathematics, Philadelphia, USA, 1979.

15. A.Neumaier: Interval Methods for Systems of Equations, Cambridge University Press, Cambridge, U.K., 1990.

16. Y.Pan and G.J.Klir: Bayesian Inference Based on Interval-Valued Prior Distributions and Likelihoods, J. of Intelligent and Fuzzy Systems, 5(1997), pp. 193-203.

17. Y.Pan and B.Yuan: Baysian Inference of Fuzzy Probabilities, Int. J. General Systems, 26(1997), pp. 73-90.

18. J.B.Paris, G.M.Wilmers and P.N.Watton: On the Structure of Probability Functions in the Natural World, Int. J. Uncertainty, Fuzziness and Knowledge-Based Systems, 8(2000), pp. 311-329.

19. J.Pykacz and B.D'Hooghe: Bell-Type Inequalities in Fuzzy Probability Calculus, Int. J. Uncertainty, Fuzziness and Knowledge-Based Systems, 9(2001), pp. 263-275.

20. F.Voorbraak: Partial Probability: Theory and Applications, Int. J. Uncertainty, Fuzziness and Knowledge-Based Systems, 8(2000), pp. 331-345.

21. P.Walley: Towards a Unified Theory of Imprecise Probability, Int. J. Approx. Reasoning, 24(2000), pp. 125-148.

22. P.Walley and G.deCooman: A Behavioral Model for Linguistic Uncertainty, Inform. Sci., 134(2001), pp. 1-37.

23. Z.Wang, K.S.Leung, M.L.Wong and J.Fang: A New Type of Nonlinear Integral and the Computational Algorithm, Fuzzy Sets and Systems, 112(2000), pp. 223-231.

24. K.Weichselberger: The Theory of Interval-Probability as a Unifying Concept for Uncertainty, Int. J. Approx. Reasoning, 24(2000), pp. 149-170.

25. L.A.Zadeh: The Concept of a Linguistic Variable and its Application to Approximate Reasoning III, Inform. Sci., 8(1975), pp. 199-249.

26. L.A.Zadeh: Fuzzy Probabilities, Information Processing and Management, 20(1984), pp. 363-372.

Chapter 2

Fuzzy Sets

2.1 Introduction

In this chapter we have collected together the basic ideas from fuzzy sets and fuzzy functions needed for the book. Any reader familiar with fuzzy sets, fuzzy numbers, the extension principle, α-cuts, interval arithmetic, and fuzzy functions may go on and have a look at Sections 2.5 through 2.10. In Section 2.5 we discuss our method of handling the maximun/minimum of a fuzzy set to be used in Chapter 9 and in Section 2.6 we present a method of ordering a finite set of fuzzy numbers from smallest to largest to be employed in Chapters 5-7. Section 2.7 will be used starting in Chapter 3 where we substitute fuzzy numbers for probabilities in discrete probability distributions. Section 2.8 is important starting in Chapter 4 where we show how to obtain fuzzy numbers for uncertain parameters in probability density (mass) functions using a set of confidence intervals. In Section 2.9 we show numerical procedures for computing α-cuts of fuzzy probabilities which will be used throughout the book. Finally, in Section 2.10, we discuss our methods of obtaining the figures for fuzzy probabilities used throughout the book. A good general reference for fuzzy sets and fuzzy logic is [4] and [17].

Our notation specifying a fuzzy set is to place a "bar" over a letter. So $\overline{A}, \overline{B}, \ldots, \overline{X}, \overline{Y}, \ldots, \overline{\alpha}, \overline{\beta}, \ldots$, will all denote fuzzy sets.

2.2 Fuzzy Sets

If Ω is some set, then a fuzzy subset $\overline{A}$ of Ω is defined by its membership function, written $\overline{A}(x)$, which produces values in $[0,1]$ for all x in Ω. So, $\overline{A}(x)$ is a function mapping Ω into $[0,1]$. If $\overline{A}(x_0) = 1$, then we say x_0 belongs to $\overline{A}$, if $\overline{A}(x_1) = 0$ we say x_1 does not belong to $\overline{A}$, and if $\overline{A}(x_2) = 0.6$ we say the membership value of x_2 in $\overline{A}$ is 0.6. When $\overline{A}(x)$ is always equal to one or zero we obtain a crisp (non–fuzzy) subset of Ω. For all fuzzy sets $\overline{B}, \overline{C}, \ldots$

we use $\overline{B}(x)$, $\overline{C}(x)$, ... for the value of their membership function at x. Most of the fuzzy sets we will be using will be fuzzy numbers .

The term "crisp" will mean not fuzzy. A crisp set is a regular set. A crisp number is just a real number. A crisp matrix (vector) has real numbers as its elements. A crisp function maps real numbers (or real vectors) into real numbers. A crisp solution to a problem is a solution involving crisp sets, crisp numbers, crisp functions, etc.

2.2.1   Fuzzy Numbers

A general definition of fuzzy number may be found in ([4],[17]), however our fuzzy numbers will be almost always triangular (shaped), or trapezoidal (shaped), fuzzy numbers. A triangular fuzzy number $\overline{N}$ is defined by three numbers $a < b < c$ where the base of the triangle is the interval $[a, c]$ and its vertex is at $x = b$. Triangular fuzzy numbers will be written as $\overline{N} = (a/b/c)$. A triangular fuzzy number $\overline{N} = (1.2/2/2.4)$ is shown in Figure 2.1. We see that $\overline{N}(2) = 1$, $\overline{N}(1.6) = 0.5$, etc.

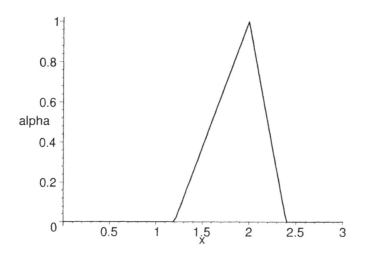

Figure 2.1: Triangular Fuzzy Number $\overline{N}$

A trapezoidal fuzzy number $\overline{M}$ is defined by four numbers $a < b < c < d$ where the base of the trapezoid is the interval $[a, d]$ and its top (where the membership equals one) is over $[b, c]$. We write $\overline{M} = (a/b, c/d)$ for trapezoidal fuzzy numbers. Figure 2.2 shows $\overline{M} = (1.2/2, 2.4/2.7)$.

A triangular shaped fuzzy number $\overline{P}$ is given in Figure 2.3. $\overline{P}$ is only partially specified by the three numbers 1.2, 2, 2.4 since the graph on [1.2, 2],

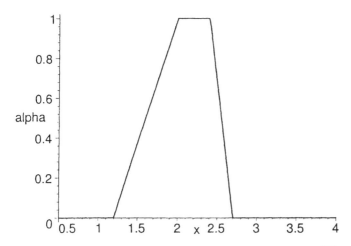

Figure 2.2: Trapezoidal Fuzzy Number $\overline{M}$

and $[2, 2.4]$, is not a straight line segment. To be a triangular shaped fuzzy number we require the graph to be continuous and: (1) monotonically increasing on $[1.2, 2]$; and (2) monotonically decreasing on $[2, 2.4]$. For triangular shaped fuzzy number $\overline{P}$ we use the notation $\overline{P} \approx (1.2/2/2.4)$ to show that it is partially defined by the three numbers 1.2, 2, and 2.4. If $\overline{P} \approx (1.2/2/2.4)$ we know its base is on the interval $[1.2, 2.4]$ with vertex (membership value one) at $x = 2$. Similarly we define trapezoidal shaped fuzzy number $\overline{Q} \approx (1.2/2, 2.4/2.7)$ whose base is $[1.2, 2.7]$ and top is over the interval $[2, 2.4]$. The graph of $\overline{Q}$ is similar to $\overline{M}$ in Figure 2.2 but it has continuous curves for its sides.

Although we will be using triangular (shaped) and trapezoidal (shaped) fuzzy numbers throughout the book, many results can be extended to more general fuzzy numbers, but we shall be content to work with only these special fuzzy numbers.

We will be using fuzzy numbers in this book to describe uncertainty. For example, in Chapter 3 a fuzzy probability can be a triangular shaped fuzzy number, it could also be a trapezoidal shaped fuzzy number. In Chapters 4 and 8-15 parameters in probability density (mass) functions, like the mean in a normal probability density function, will be a triangular fuzzy number.

2.2.2 Alpha–Cuts

Alpha–cuts are slices through a fuzzy set producing regular (non-fuzzy) sets. If $\overline{A}$ is a fuzzy subset of some set Ω, then an α–cut of $\overline{A}$, written $\overline{A}[\alpha]$ is

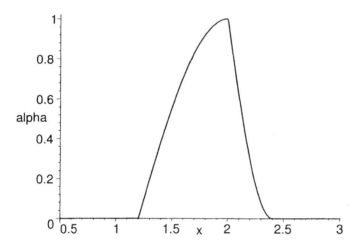

Figure 2.3: Triangular Shaped Fuzzy Number $\overline{P}$

defined as

$$\overline{A}[\alpha] = \{x \in \Omega | \overline{A}(x) \geq \alpha\} \ , \tag{2.1}$$

for all α, $0 < \alpha \leq 1$. The $\alpha = 0$ cut, or $\overline{A}[0]$, must be defined separately.

Let $\overline{N}$ be the fuzzy number in Figure 2.1. Then $\overline{N}[0] = [1.2, 2.4]$. Notice that using equation (2.1) to define $\overline{N}[0]$ would give $\overline{N}[0] =$ all the real numbers. Similarly, $\overline{M}[0] = [1.2, 2.7]$ from Figure 2.2 and in Figure 2.3 $\overline{P}[0] = [1.2, 2.4]$. For any fuzzy set $\overline{A}$, $\overline{A}[0]$ is called the support, or base, of $\overline{A}$. Many authors call the support of a fuzzy number the open interval (a, b) like the support of $\overline{N}$ in Figure 2.1 would then be $(1.2, 2.4)$. However in this book we use the closed interval $[a, b]$ for the support (base) of the fuzzy number.

The core of a fuzzy number is the set of values where the membership value equals one. If $\overline{N} = (a/b/c)$, or $\overline{N} \approx (a/b/c)$, then the core of $\overline{N}$ is the single point b. However, if $\overline{M} = (a/b, c/d)$, or $\overline{M} \approx (a/b, c/d)$, then the core of $\overline{M} = [b, c]$.

For any fuzzy number $\overline{Q}$ we know that $\overline{Q}[\alpha]$ is a closed, bounded, interval for $0 \leq \alpha \leq 1$. We will write this as

$$\overline{Q}[\alpha] = [q_1(\alpha), q_2(\alpha)] \ , \tag{2.2}$$

where $q_1(\alpha)$ $(q_2(\alpha))$ will be an increasing (decreasing) function of α with $q_1(1) \leq q_2(1)$. If $\overline{Q}$ is a triangular shaped or a trapezoidal shaped fuzzy number then: (1) $q_1(\alpha)$ will be a continuous, monotonically increasing function of α in $[0, 1]$; (2) $q_2(\alpha)$ will be a continuous, monotonically decreasing function of α, $0 \leq \alpha \leq 1$; and (3) $q_1(1) = q_2(1)$ $(q_1(1) < q_2(1)$ for trape-

zoids). We sometimes check monotone increasing (decreasing) by showing
that $dq_1(\alpha)/d\alpha > 0$ $(dq_2(\alpha)/d\alpha < 0)$ holds.

For the $\overline{N}$ in Figure 2.1 we obtain $\overline{N}[\alpha] = [n_1(\alpha), n_2(\alpha)]$, $n_1(\alpha) = 1.2 +$
0.8α and $n_2(\alpha) = 2.4 - 0.4\alpha$, $0 \leq \alpha \leq 1$. Similarly, $\overline{M}$ in Figure 2.2 has
$\overline{M}[\alpha] = [m_1(\alpha), m_2(\alpha)]$, $m_1(\alpha) = 1.2 + 0.8\alpha$ and $m_2(\alpha) = 2.7 - 0.3\alpha$, $0 \leq$
$\alpha \leq 1$. The equations for $n_i(\alpha)$ and $m_i(\alpha)$ are backwards. With the y–axis
vertical and the x–axis horizontal the equation $n_1(\alpha) = 1.2 + 0.8\alpha$ means
$x = 1.2 + 0.8y$, $0 \leq y \leq 1$. That is, the straight line segment from $(1.2, 0)$ to
$(2, 1)$ in Figure 2.1 is given as x a function of y whereas it is usually stated as
y a function of x. This is how it will be done for all α–cuts of fuzzy numbers.

2.2.3 Inequalities

Let $\overline{N} = (a/b/c)$. We write $\overline{N} \geq \delta$, δ some real number, if $a \geq \delta$, $\overline{N} > \delta$
when $a > \delta$, $\overline{N} \leq \delta$ for $c \leq \delta$ and $\overline{N} < \delta$ if $c < \delta$. We use the same notation
for triangular shaped and trapezoidal (shaped) fuzzy numbers whose support
is the interval $[a, c]$.

If $\overline{A}$ and $\overline{B}$ are two fuzzy subsets of a set Ω, then $\overline{A} \leq \overline{B}$ means $\overline{A}(x) \leq$
$\overline{B}(x)$ for all x in Ω, or $\overline{A}$ is a fuzzy subset of $\overline{B}$. $\overline{A} < \overline{B}$ holds when $\overline{A}(x) <$
$\overline{B}(x)$, for all x. There is a potential problem with the symbol $\leq$. In some
places in the book , for example see Section 2.6 and in Chapters 5-7, $\overline{M} \leq \overline{N}$,
for fuzzy numbers $\overline{M}$ and $\overline{N}$, means that $\overline{M}$ is less than or equal to $\overline{N}$. It
should be clear on how we use "$\leq$" as to which meaning is correct.

2.2.4 Discrete Fuzzy Sets

Let $\overline{A}$ be a fuzzy subset of Ω. If $\overline{A}(x)$ is not zero only at a finite number of
x values in Ω, then $\overline{A}$ is called a discrete fuzzy set. Suppose $\overline{A}(x)$ is not zero
only at x_1, x_2, x_3 and x_4 in Ω. Then we write the fuzzy set as

$$\overline{A} = \{\frac{\mu_1}{x_1}, \cdots, \frac{\mu_4}{x_4}\}, \tag{2.3}$$

where the μ_i are the membership values. That is, $\overline{A}(x_i) = \mu_i$, $1 \leq i \leq 4$,
and $\overline{A}(x) = 0$ otherwise. We can have discrete fuzzy subsets of any space Ω.
Notice that α-cuts of discrete fuzzy sets of $\mathbb{R}$, the set of real numbers, do not
produce closed, bounded, intervals.

2.3 Fuzzy Arithmetic

If $\overline{A}$ and $\overline{B}$ are two fuzzy numbers we will need to add, subtract, multiply and
divide them. There are two basic methods of computing $\overline{A} + \overline{B}$, $\overline{A} - \overline{B}$, etc.
which are: (1) extension principle; and (2) α–cuts and interval arithmetic.

2.3.1   Extension Principle

Let $\overline{A}$ and $\overline{B}$ be two fuzzy numbers. If $\overline{A} + \overline{B} = \overline{C}$, then the membership function for $\overline{C}$ is defined as

$$\overline{C}(z) = \sup_{x,y}\{\min(\overline{A}(x),\overline{B}(y))|x+y=z\} \ . \tag{2.4}$$

If we set $\overline{C} = \overline{A} - \overline{B}$, then

$$\overline{C}(z) = \sup_{x,y}\{\min(\overline{A}(x),\overline{B}(y))|x-y=z\} \ . \tag{2.5}$$

Similarly, $\overline{C} = \overline{A} \cdot \overline{B}$, then

$$\overline{C}(z) = \sup_{x,y}\{\min(\overline{A}(x),\overline{B}(y))|x\cdot y=z\} \ , \tag{2.6}$$

and if $\overline{C} = \overline{A}/\overline{B}$,

$$\overline{C}(z) = \sup_{x,y}\{\min(\overline{A}(x),\overline{B}(y))|x/y=z\} \ . \tag{2.7}$$

In all cases $\overline{C}$ is also a fuzzy number [17]. We assume that zero does not belong to the support of $\overline{B}$ in $\overline{C} = \overline{A}/\overline{B}$. If $\overline{A}$ and $\overline{B}$ are triangular (trapezoidal) fuzzy numbers then so are $\overline{A} + \overline{B}$ and $\overline{A} - \overline{B}$, but $\overline{A} \cdot \overline{B}$ and $\overline{A}/\overline{B}$ will be triangular (trapezoidal) shaped fuzzy numbers.

We should mention something about the operator "sup" in equations (2.4) – (2.7). If Ω is a set of real numbers bounded above (there is a M so that $x \leq M$, for all x in Ω), then $\sup(\Omega)$ = the least upper bound for Ω. If Ω has a maximum member, then $\sup(\Omega) = \max(\Omega)$. For example, if $\Omega = [0,1)$, $\sup(\Omega) = 1$ but if $\Omega = [0,1]$, then $\sup(\Omega) = \max(\Omega) = 1$. The dual operator to "sup" is "inf". If Ω is bounded below (there is a M so that $M \leq x$ for all $x \in \Omega$), then $\inf(\Omega)$ = the greatest lower bound. For example, for $\Omega = (0,1]$ $\inf(\Omega) = 0$ but if $\Omega = [0,1]$, then $\inf(\Omega) = \min(\Omega) = 0$.

Obviously, given $\overline{A}$ and $\overline{B}$, equations (2.4) – (2.7) appear quite complicated to compute $\overline{A} + \overline{B}$, $\overline{A} - \overline{B}$, etc. So, we now present an equivalent procedure based on α–cuts and interval arithmetic. First, we present the basics of interval arithmetic.

2.3.2   Interval Arithmetic

We only give a brief introduction to interval arithmetic. For more information the reader is referred to ([19],[20]). Let $[a_1, b_1]$ and $[a_2, b_2]$ be two closed, bounded, intervals of real numbers. If $*$ denotes addition, subtraction, multiplication, or division, then $[a_1, b_1] * [a_2, b_2] = [\alpha, \beta]$ where

$$[\alpha,\beta] = \{a * b | a_1 \leq a \leq b_1, a_2 \leq b \leq b_2\} \ . \tag{2.8}$$

If $*$ is division, we must assume that zero does not belong to $[a_2, b_2]$. We may simplify equation (2.8) as follows:

$$[a_1, b_1] + [a_2, b_2] = [a_1 + a_2, b_1 + b_2] , \qquad (2.9)$$

$$[a_1, b_1] - [a_2, b_2] = [a_1 - b_2, b_1 - a_2] , \qquad (2.10)$$

$$[a_1, b_1] / [a_2, b_2] = [a_1, b_1] \cdot \left[\frac{1}{b_2}, \frac{1}{a_2} \right] , \qquad (2.11)$$

and

$$[a_1, b_1] \cdot [a_2, b_2] = [\alpha, \beta] , \qquad (2.12)$$

where

$$\alpha = \min\{a_1 a_2, a_1 b_2, b_1 a_2, b_1 b_2\} , \qquad (2.13)$$

$$\beta = \max\{a_1 a_2, a_1 b_2, b_1 a_2, b_1 b_2\} . \qquad (2.14)$$

Multiplication and division may be further simplified if we know that $a_1 > 0$ and $b_2 < 0$, or $b_1 > 0$ and $b_2 < 0$, etc. For example, if $a_1 \geq 0$ and $a_2 \geq 0$, then

$$[a_1, b_1] \cdot [a_2, b_2] = [a_1 a_2, b_1 b_2] , \qquad (2.15)$$

and if $b_1 < 0$ but $a_2 \geq 0$, we see that

$$[a_1, b_1] \cdot [a_2, b_2] = [a_1 b_2, a_2 b_1] . \qquad (2.16)$$

Also, assuming $b_1 < 0$ and $b_2 < 0$ we get

$$[a_1, b_1] \cdot [a_2, b_2] = [b_1 b_2, a_1 a_2] , \qquad (2.17)$$

but $a_1 \geq 0$, $b_2 < 0$ produces

$$[a_1, b_1] \cdot [a_2, b_2] = [a_2 b_1, b_2 a_1] . \qquad (2.18)$$

2.3.3 Fuzzy Arithmetic

Again we have two fuzzy numbers $\overline{A}$ and $\overline{B}$. We know α–cuts are closed, bounded, intervals so let $\overline{A}[\alpha] = [a_1(\alpha), a_2(\alpha)]$, $\overline{B}[\alpha] = [b_1(\alpha), b_2(\alpha)]$. Then if $\overline{C} = \overline{A} + \overline{B}$ we have

$$\overline{C}[\alpha] = \overline{A}[\alpha] + \overline{B}[\alpha] . \qquad (2.19)$$

We add the intervals using equation (2.9). Setting $\overline{C} = \overline{A} - \overline{B}$ we get

$$\overline{C}[\alpha] = \overline{A}[\alpha] - \overline{B}[\alpha] , \qquad (2.20)$$

for all α in $[0, 1]$. Also

$$\overline{C}[\alpha] = \overline{A}[\alpha] \cdot \overline{B}[\alpha] , \qquad (2.21)$$

for $\overline{C} = \overline{A} \cdot \overline{B}$ and

$$\overline{C}[\alpha] = \overline{A}[\alpha] / \overline{B}[\alpha] , \qquad (2.22)$$

when $\overline{C} = \overline{A}/\overline{B}$, provided that zero does not belong to $\overline{B}[\alpha]$ for all α. This method is equivalent to the extension principle method of fuzzy arithmetic [17]. Obviously, this procedure, of α–cuts plus interval arithmetic, is more user (and computer) friendly.

Example 2.3.3.1

Let $\overline{A} = (-3/-2/-1)$ and $\overline{B} = (4/5/6)$. We determine $\overline{A} \cdot \overline{B}$ using α–cuts and interval arithmetic. We compute $\overline{A}[\alpha] = [-3 + \alpha, -1 - \alpha]$ and $\overline{B}[\alpha] = [4+\alpha, 6-\alpha]$. So, if $\overline{C} = \overline{A} \cdot \overline{B}$ we obtain $\overline{C}[\alpha] = [(\alpha-3)(6-\alpha), (-1-\alpha)(4+\alpha)]$, $0 \leq \alpha \leq 1$. The graph of $\overline{C}$ is shown in Figure 2.4.

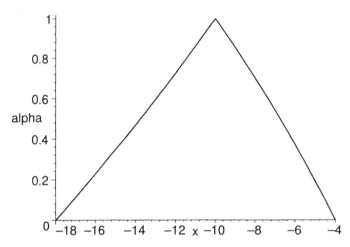

Figure 2.4: The Fuzzy Number $\overline{C} = \overline{A} \cdot \overline{B}$

2.4 Fuzzy Functions

In this book a fuzzy function is a mapping from fuzzy numbers into fuzzy numbers. We write $H(\overline{X}) = \overline{Z}$ for a fuzzy function with one independent variable $\overline{X}$. Usually $\overline{X}$ will be a triangular (trapezoidal) fuzzy number and then we usually obtain $\overline{Z}$ as a triangular (trapezoidal) shaped fuzzy number. For two independent variables we have $H(\overline{X}, \overline{Y}) = \overline{Z}$.

Where do these fuzzy functions come from? They are usually extensions of real–valued functions. Let $h : [a, b] \to \mathbb{R}$. This notation means $z = h(x)$ for x in $[a, b]$ and z a real number. One extends $h : [a, b] \to \mathbb{R}$ to $H(\overline{X}) = \overline{Z}$ in two ways: (1) the extension principle; or (2) using α–cuts and interval arithmetic.

2.4.1 Extension Principle

Any $h : [a, b] \to \mathbb{R}$ may be extended to $H(\overline{X}) = \overline{Z}$ as follows

$$\overline{Z}(z) = \sup_{x} \left\{ \overline{X}(x) \mid h(x) = z, \ a \leq x \leq b \right\} . \tag{2.23}$$

Equation (2.23) defines the membership function of $\overline{Z}$ for any triangular (trapezoidal) fuzzy number $\overline{X}$ in $[a, b]$.

If h is continuous, then we have a way to find α–cuts of $\overline{Z}$. Let $\overline{Z}[\alpha] = [z_1(\alpha), z_2(\alpha)]$. Then [8]

$$z_1(\alpha) \;=\; \min\{\, h(x) \mid x \in \overline{X}[\alpha] \,\}\,, \qquad (2.24)$$
$$z_2(\alpha) \;=\; \max\{\, h(x) \mid x \in \overline{X}[\alpha] \,\}\,, \qquad (2.25)$$

for $0 \le \alpha \le 1$.

If we have two independent variables, then let $z = h(x, y)$ for x in $[a_1, b_1]$, y in $[a_2, b_2]$. We extend h to $H(\overline{X}, \overline{Y}) = \overline{Z}$ as

$$\overline{Z}(z) = \sup_{x,y} \left\{ \min\left(\overline{X}(x), \overline{Y}(y)\right) \mid h(x, y) = z \right\}\,, \qquad (2.26)$$

for $\overline{X}$ $(\overline{Y})$ a triangular or trapezoidal fuzzy number in $[a_1, b_1]$ $([a_2, b_2])$. For α–cuts of $\overline{Z}$, assuming h is continuous, we have

$$z_1(\alpha) \;=\; \min\{\, h(x, y) \mid x \in \overline{X}[\alpha],\ y \in \overline{Y}[\alpha] \,\}\,, \qquad (2.27)$$
$$z_2(\alpha) \;=\; \max\{\, h(x, y) \mid x \in \overline{X}[\alpha],\ y \in \overline{Y}[\alpha] \,\}\,, \qquad (2.28)$$

$0 \le \alpha \le 1$. We use equations (2.24) – (2.25) and (2.27) – (2.28) throughout this book.

Applications

Let $f(x_1, ..., x_n; \theta_1, ..., \theta_m)$ be a continuous function . Then

$$I[\alpha] = \{f(x_1, ..., x_n; \theta_1, ..., \theta_m) \mid \;\; \mathbf{S} \;\; \}, \qquad (2.29)$$

for $\alpha \in [0, 1]$ and $\mathbf{S}$ is the statement "$\theta_i \in \overline{\theta}_i[\alpha]$, $1 \le i \le m$", for fuzzy numbers $\overline{\theta}_i$, $1 \le i \le m$, defines an interval $I[\alpha]$. The endpoints of $I[\alpha]$ may be found as in equations (2.24),(2.25) and (2.27),(2.28). $I[\alpha]$ gives the α-cuts of $f(x_1, ..., x_n; \overline{\theta}_i, ..., \overline{\theta}_m)$.

We may also reverse the above procedure. Let $h(x_1, ..., x_n; \tau_1, ..., \tau_m)$ be a continuous function. Define

$$\Gamma[\alpha] = \{h(x_1, ..., x_n; \tau_1, ..., \tau_m) \mid \;\; \mathbf{S} \;\; \}, \qquad (2.30)$$

for $\alpha \in [0, 1]$, $\mathbf{S}$ is "$\tau_i \in \overline{\tau}_i[\alpha]$, $1 \le i \le m$" and the $\overline{\tau}_i$, $1 \le i \le m$ are fuzzy numbers. Then the $\Gamma[\alpha]$ are intervals giving the α-cuts of fuzzy function $h(x_1, ..., x_n; \overline{\tau}_1, ..., \overline{\tau}_m)$.

These two results will be used throughout the book.

2.4.2   Alpha–Cuts and Interval Arithmetic

All the functions we usually use in engineering and science have a computer algorithm which, using a finite number of additions, subtractions, multiplications and divisions, can evaluate the function to required accuracy [7]. Such functions can be extended, using α–cuts and interval arithmetic, to fuzzy functions. Let $h : [a, b] \to \mathbb{R}$ be such a function. Then its extension $H(\overline{X}) = \overline{Z}$, $\overline{X}$ in $[a, b]$ is done, via interval arithmetic, in computing $h(\overline{X}[\alpha]) = \overline{Z}[\alpha]$, α in $[0, 1]$. We input the interval $\overline{X}[\alpha]$, perform the arithmetic operations needed to evaluate h on this interval, and obtain the interval $\overline{Z}[\alpha]$. Then put these α–cuts together to obtain the value $\overline{Z}$. The extension to more independent variables is straightforward.

For example, consider the fuzzy function

$$\overline{Z} = H(\overline{X}) = \frac{\overline{A}\,\overline{X} + \overline{B}}{\overline{C}\,\overline{X} + \overline{D}} \; , \tag{2.31}$$

for triangular fuzzy numbers $\overline{A}, \overline{B}, \overline{C}, \overline{D}$ and triangular fuzzy number $\overline{X}$ in $[0, 10]$. We assume that $\overline{C} \geq 0$, $\overline{D} > 0$ so that $\overline{C}\,\overline{X} + \overline{D} > 0$. This would be the extension of

$$h(x_1, x_2, x_3, x_4, x) = \frac{x_1 x + x_2}{x_3 x + x_4} \; . \tag{2.32}$$

We would substitute the intervals $\overline{A}[\alpha]$ for x_1, $\overline{B}[\alpha]$ for x_2, $\overline{C}[\alpha]$ for x_3, $\overline{D}[\alpha]$ for x_4 and $\overline{X}[\alpha]$ for x, do interval arithmetic, to obtain interval $\overline{Z}[\alpha]$ for $\overline{Z}$. Alternatively, the fuzzy function

$$\overline{Z} = H(\overline{X}) = \frac{2\overline{X} + 10}{3\overline{X} + 4} \; , \tag{2.33}$$

would be the extension of

$$h(x) = \frac{2x + 10}{3x + 4} \; . \tag{2.34}$$

2.4.3   Differences

Let $h : [a, b] \to \mathbb{R}$. Just for this subsection let us write $\overline{Z}^* = H(\overline{X})$ for the extension principle method of extending h to H for $\overline{X}$ in $[a, b]$. We denote $\overline{Z} = H(\overline{X})$ for the α–cut and interval arithmetic extension of h .

We know that $\overline{Z}$ can be different from $\overline{Z}^*$. But for basic fuzzy arithmetic in Section 2.2 the two methods give the same results. In the example below we show that for $h(x) = x(1 - x)$, x in $[0, 1]$, we can get $\overline{Z}^* \neq \overline{Z}$ for some $\overline{X}$ in $[0, 1]$. What is known ([8],[19]) is that for usual functions in science and engineering $\overline{Z}^* \leq \overline{Z}$. Otherwise, there is no known necessary and sufficient conditions on h so that $\overline{Z}^* = \overline{Z}$ for all $\overline{X}$ in $[a, b]$.

There is nothing wrong in using α–cuts and interval arithmetic to evaluate fuzzy functions. Surely, it is user, and computer friendly. However, we should be aware that whenever we use α–cuts plus interval arithmetic to compute $\overline{Z} = H(\overline{X})$ we may be getting something larger than that obtained from the extension principle. The same results hold for functions of two or more independent variables.

Example 2.4.3.1

The example is the simple fuzzy expression

$$\overline{Z} = (1 - \overline{X})\ \overline{X}\ , \tag{2.35}$$

for $\overline{X}$ a triangular fuzzy number in $[0, 1]$. Let $\overline{X}[\alpha] = [x_1(\alpha), x_2(\alpha)]$. Using interval arithmetic we obtain

$$
\begin{aligned}
z_1(\alpha) &= (1 - x_2(\alpha))x_1(\alpha)\ , & (2.36) \\
z_2(\alpha) &= (1 - x_1(\alpha))x_2(\alpha)\ , & (2.37)
\end{aligned}
$$

for $\overline{Z}[\alpha] = [z_1(\alpha), z_2(\alpha)]$, α in $[0, 1]$.

The extension principle extends the regular equation $z = (1 - x)x$, $0 \leq x \leq 1$, to fuzzy numbers as follows

$$\overline{Z}^*(z) = \sup_x \left\{ \overline{X}(x) | (1 - x)x = z,\ 0 \leq x \leq 1 \right\}\ . \tag{2.38}$$

Let $\overline{Z}^*[\alpha] = [z_1^*(\alpha), z_2^*(\alpha)]$. Then

$$
\begin{aligned}
z_1^*(\alpha) &= \min\{(1 - x)x | x \in \overline{X}[\alpha]\}\ , & (2.39) \\
z_2^*(\alpha) &= \max\{(1 - x)x | x \in \overline{X}[\alpha]\}\ , & (2.40)
\end{aligned}
$$

for all $0 \leq \alpha \leq 1$. Now let $\overline{X} = (0/0.25/0.5)$, then $x_1(\alpha) = 0.25\alpha$ and $x_2(\alpha) = 0.50 - 0.25\alpha$. Equations (2.36) and (2.37) give $\overline{Z}[0.50] = [5/64, 21/64]$ but equations (2.39) and (2.40) produce $\overline{Z}^*[0.50] = [7/64, 15/64]$. Therefore, $\overline{Z}^* \neq \overline{Z}$. We do know that if each fuzzy number appears only once in the fuzzy expression, the two methods produce the same results ([8],[19]). However, if a fuzzy number is used more than once, as in equation (2.35), the two procedures can give different results.

2.5 Finding the Minimum of a Fuzzy Number

In Chapter 9 we will want to determine the values of some decision variables $y = (x_1, ..., x_n)$ that will minimize a fuzzy function $\overline{E}(y)$. For each value of y we obtain a fuzzy number $\overline{E}(y)$.

We can not minimize a fuzzy number so what we are going to do, which we have done before ([6],[9]-[13]), is first change $min\overline{E}(y)$ into a multiobjective problem and then translate the multiobjective problem into a single objective problem. This strategy is adopted from the finance literature where they had the problem of minimizing a random variable X whose values are constrained by a probability density function $g(x)$. They considered the multiobjective problem: (1) minimize the expected value of X; (2) minimize the variance of X; and (3) minimize the skewness of X to the right of the expected value. For our problem let: (1) $c(y)$ be the center of the core of $\overline{E}(y)$, the core of a fuzzy number is the interval where the membership function equals one, for each y; (2) $L(y)$ be the area under the graph of the membership function to the left of $c(y)$; and (3) $R(y)$ be the area under the graph of the membership function to the right of $c(y)$. See Figure 2.5. For $min\overline{E}(y)$ we substitute: (1) $min[c(y)]$; (2) $maxL(y)$, or maximize the possibility of obtaining values less than $c(y)$; and (3) $minR(y)$, or minimize the possibility of obtaining values greater then $c(y)$. So for $min\overline{E}(y)$ we have

$$V = (maxL(y), min[c(y)], minR(y)).\qquad (2.41)$$

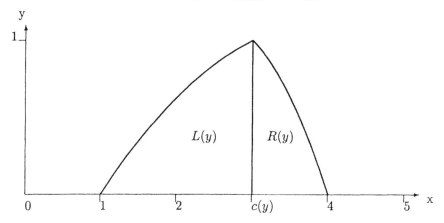

Figure 2.5: Computations for the Min. of a Fuzzy Number

First let M be a sufficiently large positive number so that $maxL(y)$ is equivalent to $minL^*(y)$ where $L^*(y) = M - L(y)$. The multiobjective problem become

$$minV' = (minL^*(y), min[c(y)], minR(y)).\qquad (2.42)$$

In a multiobjective optimization problem a solution is a value of the decision variable y that produces an undominated vector V'. Let V be the set of all vectors V' obtained for all possible values of the decision variable y. Vector $v_a = (v_{a1}, v_{a2}, v_{a3})$ dominates vector $v_b = (v_{b1}, v_{b2}, v_{b3})$, both in V, if $v_{ai} \leq v_{bi}$, $1 \leq i \leq 3$, with one of the $\leq$ a strict inequality $<$. A vector $v \in V$ is undominated if no $w \in V$ dominates v. The set of undominated vectors in V is considered the general solution and the problem is to find values of the decision variables that produce undominated V'. The above definition of undominated was for a min problem, obvious changes need to be made for a max problem.

One way to explore the undominated set is to change the multiobjective problem into a single objective. The single objective problem is

$$min(\lambda_1[M - L(y)] + \lambda_2 c(y) + \lambda_3 R(y)), \qquad (2.43)$$

where $\lambda_i > 0$, $1 \leq i \leq 3$, $\lambda_1 + \lambda_2 + \lambda_3 = 1$. You will get different undominated solutions by choosing different values of $\lambda_i > 0$, $\lambda_1 + \lambda_2 + \lambda_3 = 1$. It is known that solutions to this problem are undominated, but for some problems it will be unable to generate all undominated solutions [16]. The decision maker is to choose the values of the weights λ_i for the three minimization goals. Usually one picks different values for the λ_i to explore the solution set and then lets the decision maker choose an optimal y^* from this set of solutions.

This is how we propose to handle the problem of $min\overline{E}(y)$ in fuzzy inventory control in Section 9.2 in Chapter 9. Numerical solutions to this optimization problem can be difficult, depending on the constraints. In the past we have employed an evolutionary algorithm to generate good approximate solutions. See ([4],[5],[9]) for a general description of our evolutionary algorithm and other applications to solving fuzzy optimization problems. Obvious changes need to be made in the above discussion for a *max* problem when we consider $max\overline{E}(y)$ in Section 9.3 in Chapter 9.

2.6 Ordering Fuzzy Numbers

Given a finite set of fuzzy numbers $\overline{A}_1, ..., \overline{A}_n$ in Chapters 5-7, we want to order them from smallest to largest. Each $\overline{A}_i$ corresponds to a decision variable a_i, $1 \leq i \leq n$ and in a *max* (*min*) problem the largest (smallest) $\overline{A}_i$ gives the optimal choice for the decision variables. For a finite set of real numbers there is no problem in ordering them from smallest to largest. However, in the fuzzy case there is no universally accepted way to do this. There are probably more than 40 methods proposed in the literature of defining $\overline{M} \leq \overline{N}$, for two fuzzy numbers $\overline{M}$ and $\overline{N}$. Here the symbol $\leq$ means "less than or equal" and not "a fuzzy subset of". A few key references on this topic are ([1],[14],[15],[22],[23]), where the interested reader can look up many of these methods and see their comparisons.

Here we will present only one procedure for ordering fuzzy numbers that we have used before ([2],[3]). But note that different definitions of $\leq$ between fuzzy numbers can give different ordering. A different procedure for defining "$\leq$" between fuzzy numbers is defined, and used, in Chapter 14. We first define $<$ between two fuzzy numbers $\overline{M}$ and $\overline{N}$. Define

$$v(\overline{M} \leq \overline{N}) = max\{min(\overline{M}(x), \overline{N}(y))|x \leq y\}, \qquad (2.44)$$

which measures how much $\overline{M}$ is less than or equal to $\overline{N}$. We write $\overline{N} < \overline{M}$ if $v(\overline{N} \leq \overline{M}) = 1$ but $v(\overline{M} \leq \overline{N}) < \eta$, where η is some fixed fraction in $(0, 1]$. In this book we will usually use $\eta = 0.8$. Then $\overline{N} < \overline{M}$ if $v(\overline{N} \leq \overline{M}) = 1$

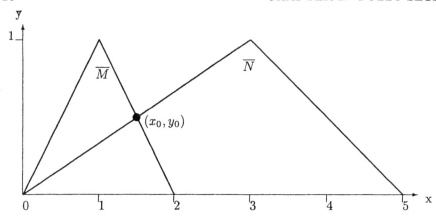

Figure 2.6: Determining $v(\overline{N} \leq \overline{M})$

and $v(\overline{M} \leq \overline{N}) < 0.8$. We then define $\overline{M} \approx \overline{N}$ when both $\overline{N} < \overline{M}$ and $\overline{M} < \overline{N}$ are false. $\overline{M} \leq \overline{N}$ means $\overline{M} < \overline{N}$ or $\overline{M} \approx \overline{N}$. Now this $\approx$ may not be transitive. If $\overline{N} \approx \overline{M}$ and $\overline{M} \approx \overline{O}$ implies that $\overline{N} \approx \overline{O}$, then $\approx$ is transitive. However, it can happen that $\overline{N} \approx \overline{M}$ and $\overline{M} \approx \overline{O}$ but $\overline{N} < \overline{O}$ because $\overline{M}$ lies a little to the right of $\overline{N}$ and $\overline{O}$ lies a little to the right of $\overline{M}$ but $\overline{O}$ lies sufficiently far to the right of $\overline{N}$ that we obtain $\overline{N} < \overline{O}$. But this ordering is still useful in partitioning the set of fuzzy numbers up into sets $H_1, ..., H_K$ where ([2],[3]): (1) Given any $\overline{M}$ and $\overline{N}$ in H_k, $1 \leq k \leq K$, then $\overline{M} \approx \overline{N}$; and (2) given $\overline{N} \in H_i$ and $\overline{M} \in H_j$, with $i < j$, then $\overline{N} < \overline{M}$. Then the highest ranked fuzzy numbers lie in H_K, the second highest ranked fuzzy numbers are in H_{K-1}, etc. This result is easily seen if you graph all the fuzzy numbers on the same axis then those in H_K will be clustered together farthest to the right, proceeding from the H_K cluster to the left the next cluster will be those in H_{K-1}, etc. Then in a *max* (*min*) decision problem the optimal values of the decision variables correspond to those fuzzy sets in H_K (H_1). If you require a unique decision, then you will need to decide between those fuzzy numbers in the highest ranked set.

There is an easy way to determine if $\overline{M} < \overline{N}$, or $\overline{M} \approx \overline{N}$, for many fuzzy numbers. First, it is easy to see that if the core of $\overline{N}$ lies completely to the right of the core of $\overline{M}$, then $v(\overline{M} \leq \overline{N}) = 1$. Also, if the core of $\overline{M}$ and the core of $\overline{N}$ overlap, then $\overline{M} \approx \overline{N}$. Now assume that the core of $\overline{N}$ lies to the right of the core of $\overline{M}$, as shown in Figure 2.6 for triangular fuzzy numbers, and we wish to compute $v(\overline{N} \leq \overline{M})$. The value of this expression is simply y_0 in Figure 2.6. In general, for triangular (shaped), and trapezoidal (shaped), fuzzy numbers $v(\overline{N} \leq \overline{M})$ is the height of their intersection when the core of $\overline{N}$ lies to the right of the core of $\overline{M}$.

2.7 Fuzzy Probabilities

Let $X = \{x_1, ..., x_n\}$ be a finite set and let P be a probability function defined on all subsets of X with $P(\{x_i\}) = a_i$, $1 \leq i \leq n$, $0 < a_i < 1$, all i, and $\sum_{i=1}^{n} a_i = 1$. Starting in Chapter 3 we will substitute a fuzzy number $\bar{a}_i$ for a_i, for some i, to obtain a discrete (finite) fuzzy probability distribution. Where do these fuzzy numbers come from?

In some problems, because of the way the problem is stated, the values of all the a_i are crisp and known. For example, consider tossing a fair coin and $a_1 =$ the probability of getting a "head" and $a_2 =$ is the probability of obtaining a "tail". Since we assumed it to be a fair coin we must have $a_1 = a_2 = 0.5$. In this case we would not substitute a fuzzy number for a_1 or a_2. But in many other problems the a_i are not known exactly and they are either estimated from a random sample or they are obtained from "expert opinion".

Suppose we have the results of a random sample to estimate the value of a_1. We would construct a set of confidence intervals for a_1 and then put these together to get the fuzzy number $\bar{a}_1$ for a_1. This method of building a fuzzy number from confidence intervals is discussed in detail in the next section.

Assume that we do not know the values of the a_i and we do not have any data to estimate their values. Then we may obtain numbers for the a_i from some group of experts. This group could consist of only one expert. This case includes subjective, or "personal", probabilities in Chapter 7.

First assume we have only one expert and he is to estimate the value of some probability p. We can solicit this estimate from the expert as is done in estimating job times in project scheduling ([21], Chapter 13). Let $a =$ the "pessimistic" value of p, or the smallest possible value, let $c =$ be the "optimistic" value of p, or the highest possible value, and let $b =$ the most likely value of p. We then ask the expert to give values for a, b, c and we construct the triangular fuzzy number $\bar{p} = (a/b/c)$ for p. If we have a group of N experts all to estimate the value of p we solicit the a_i, b_i and c_i, $1 \leq i \leq N$, from them. Let a be the average of the a_i, b is the mean of the b_i and c is the average of the c_i. The simplest thing to do is to use $(a/b/c)$ for p.

2.8 Fuzzy Numbers from Confidence Intervals

We will be using fuzzy numbers for parameters in probability density functions (probability mass functions, the discrete case) beginning in Chapter 4 and in this section we show how we obtain these fuzzy numbers from a set of confidence intervals . Let X be a random variable with probability density function (or probability mass function) $f(x; \theta)$ for single parameter θ. It is easy to generalize our method to the case where θ is a vector of parameters.

Assume that θ is unknown and it must be estimated from a random sample $X_1, ..., X_n$. Let $Y = u(X_1, ..., X_n)$ be a statistic used to estimate θ. Given the values of these random variables $X_i = x_i$, $1 \leq i \leq n$, we obtain a point estimate $\theta^* = y = u(x_1, ..., x_n)$ for θ. We would never expect this point estimate to exactly equal θ so we often also compute a $(1 - \beta)100\%$ confidence interval for θ. We are using β here since α, usually employed for confidence interval, is reserved for α-cuts of fuzzy numbers. In this confidence interval one usually sets β equal to 0.10, 0.05 or 0.01.

We propose to find the $(1 - \beta)100\%$ confidence interval for all $0.01 \leq \beta < 1$. Starting at 0.01 is arbitrary and you could begin at 0.001 or 0.005 etc. Denote these confidence intervals as

$$[\theta_1(\beta), \theta_2(\beta)], \tag{2.45}$$

for $0.01 \leq \beta < 1$. Add to this the interval $[\theta^*, \theta^*]$ for the 0% confidence interval for θ. Then we have $(1 - \beta)100\%$ confidence interval for θ for $0.01 \leq \beta \leq 1$.

Now place these confidence intervals, one on top of the other, to produce a triangular shaped fuzzy number $\overline{\theta}$ whose α-cuts are the confidence intervals. We have

$$\overline{\theta}[\alpha] = [\theta_1(\alpha), \theta_2(\alpha)], \tag{2.46}$$

for $0.01 \leq \alpha \leq 1$. All that is needed is to finish the "bottom" of $\overline{\theta}$ to make it a complete fuzzy number. We will simply drop the graph of $\overline{\theta}$ straight down to complete its α-cuts so

$$\overline{\theta}[\alpha] = [\theta_1(0.01), \theta_2(0.01)], \tag{2.47}$$

for $0 \leq \alpha < 0.01$. In this way we are using more information in $\overline{\theta}$ than just a point estimate, or just a single interval estimate.

The following example shows that the fuzzy mean of the normal probability density will be a triangular shaped fuzzy number. However, for simplicity, throughout this book we will always use triangular fuzzy numbers for the fuzzy values of uncertain parameters in probability density (mass) functions.

Example 2.8.1

Consider X a random variable with probability density function $N(\mu, 100)$, which is the normal probability density with unknown mean μ and known variance $\sigma^2 = 100$. To estimate μ we obtain a random sample $X_1, ..., X_n$ from $N(\mu, 100)$. Suppose the mean of this random sample turns out to be 28.6. Then a $(1 - \beta)100\%$ confidence interval for μ is

$$[\theta_1(\beta), \theta_2(\beta)] = [28.6 - z_{\beta/2}10/\sqrt{n}, 28.6 + z_{\beta/2}10/\sqrt{n}], \tag{2.48}$$

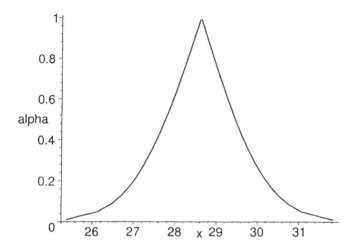

Figure 2.7: Fuzzy Mean $\overline{\mu}$ in Example 2.7.1

where $z_{\beta/2}$ is defined as

$$\int_{-\infty}^{z_{\beta/2}} N(0,1)dx = 1 - \beta/2, \tag{2.49}$$

and $N(0,1)$ denotes the normal density with mean zero and unit variance. To obtain a graph of fuzzy μ, or $\overline{\mu}$, let $n = 64$. Then we evaluated equations (2.48) and (2.49) using Maple [18] and then the final graph of $\overline{\mu}$ is shown in Figure 2.7, without dropping the graph straight down to the x-axis at the end points.

In future chapters we will have fuzzy numbers for the parameters in the probability density (mass) functions producing fuzzy probability density (mass) functions and this may be justified by the discussion presented above.

2.9   Computing Fuzzy Probabilities

Throughout this book whenever we wish to find the α-cut of a fuzzy probability we will need to solve an optimization problem. The problem is to find the max and min of a function $f(p_1, ..., p_n)$ subject to linear constraints. There will be two types of problems. The first one, needed beginning in Chapter 3 will be described in the next subsection and the other type of problem, used starting in Chapter 4, is discussed in the second subsection.

2.9.1   First Problem

The structure of this problem is

$$max/min f(p_{i_1}, ..., p_{i_K})\tag{2.50}$$

subject to

$$a_i \le p_i \le b_i, 1 \le i \le n,\tag{2.51}$$

and

$$p_1 + ... + p_n = 1.\tag{2.52}$$

The set $\{p_{i_1}, ..., p_{i_K}\}$ is a subset of $\{p_1, ..., p_n\}$. The p_i must be in interval $[a_i, b_i]$, $1 \le i \le n$ and their sum must be one. In the application of this problem : (1) the p_i will be probabilities ; (2) the intervals $[a_i, b_i]$ will be α-cuts of fuzzy numbers used for fuzzy probabilities; and (3) the sum of the p_i equals one means the sum of the probabilities is one. This problem is actually solving for the α-cuts of a fuzzy probability.

In this subsection we explain how we will obtain numerical solutions to this problem. If $f(p_{i_1}, ..., p_{i_K})$ is a linear function of the $p_i's$, then this problem is a linear programming problem and we can solve it using the "simplex" call in Maple [18]. For example, "simplex" was used to solve these problems in Example 7.2.2 in Chapter 7. So now assume that the function f is not a linear function of the $p_i's$. We used two methods of solution: (1)graphical; and (2) calculus. The calculus procedure is discussed in the three examples below, so now let us discuss the graphical method.

The graphical method is applicable for $n = 2$, and sometimes for $n = 3$. First let $n = 2$ and assume that $\bar{p}_2 = 1 - \bar{p}_1$ so that we may substitute $1 - p_1$ for p_2 in the function f and obtain f a function of p_1 only. The optimization problem is now $max/min f(p_1)$ subject to $a_1 \le p_1 \le b_1$. Assuming the calculus method discussed below is not applicable, we simply used Maple to graph $f(p_1)$ for $a_1 \le p_1 \le b_1$. From the graph we can sometimes easily find the max, or min, especially if they are at the end points. Now suppose the max (min) is not at an end point. Then we repeatedly evaluated the function in the neighborhood of the extreme point until we could estimate its value to our desired accuracy. Next let $n = 3$ and assume that $\bar{p}_3 = 1 - (\bar{p}_1 + \bar{p}_2)$. For example $\bar{p}_1 = (0.2/0.3/0.4) = \bar{p}_2$ and $\bar{p}_3 = (0.2/0.4/0.6)$ satisfies this constraint. Then substitute $1 - (p_1 + p_2)$ for p_3 in f to get f a function of only p_1 and p_2. Then the optimization problem becomes $max/min f(p_1, p_2)$ subject to $a_1 \le p_1 \le b_1$, $a_2 \le p_2 \le b_2$. We used Maple to graph the surface over the rectangle $a_i \le p_i \le b_i$, $i = 1, 2$. Then, as in the $n = 2$ case, we found by inspection extreme points, or we used Maple to repeatedly evaluate the function in the neighborhood of an extreme point to estimate the max, or min, value. This method of looking at the surface $z = f(p_1, p_2)$ is also applicable when $n = 2$ and we can not use $\bar{p}_2 = 1 - \bar{p}_1$.

Now let us look at three examples of the calculus method. But first we need to define what we will mean by saying that a certain subset of the

p_i, $1 \leq i \leq n$, is feasible. To keep the notation simple let $n = 5$ and we claim that p_1, p_2 and p_4 are feasible. This means that we may choose any $p_i \in \bar{p}_i[\alpha]$, $i = 1, 2, 4$, and then we can then find a $p_3 \in \bar{p}_3[\alpha]$ and a $p_5 \in \bar{p}_5[\alpha]$ so that $p_1 + p_2 + p_3 + p_4 + p_5 = 1$. Let $\bar{p}_i[\alpha] = [p_{i1}(\alpha), p_{i2}(\alpha)]$, $i = 1, ..., 5$ and $0 \leq \alpha \leq 1$ and assume that f is a function of only p_1, p_2 and p_4. Also assume that f is: (1) an increasing function of p_1 and p_4; and (2) a decreasing function of p_2. We will be using $[a_i, b_i] = \bar{p}_i[\alpha]$, $1 \leq i \leq 5$. If p_1, p_2 and p_4 are feasible, then, as in the examples below, we may find that for $\alpha \in [0, 1]$: (1) $minf(p_1, p_2, p_4) = f(p_{11}(\alpha), p_{22}(\alpha), p_{41}(\alpha))$; and (2)$maxf(p_1, p_2, p_4) = f(p_{12}(\alpha), p_{21}(\alpha), p_{42}(\alpha))$.

Example 2.9.1.1

Consider the problem

$$max/minf(p_1, p_2) \qquad (2.53)$$

subject to

$$p_1 \in [a_1, b_1], p_2 \in [a_2, b_2], p_1 + p_2 = 1, \qquad (2.54)$$

where $0 \leq a_i \leq b_i \leq 1$, $i = 1, 2$. Also assume that $\partial f / \partial p_1 > 0$ and $\partial f / \partial p_2 < 0$ for the p_i in $[0, 1]$. Now the result depends on the two intervals $[a_1, b_1]$ and $[a_2, b_2]$.

First assume that $1 - a_1 = b_2$ and $1 - b_1 = a_2$. This means that $p_1 = a_1$ and $p_2 = b_2$ are feasible because $a_1 + b_2 = 1$. Also, $p_1 = b_1$ and $p_2 = a_2$ are feasible since $b_1 + a_2 = 1$. For example, $[0.3, 0.6]$ and $[0.4, 0.7]$ are two such intervals. Then

$$minf(p_1, p_2) = f(a_1, b_2), \qquad (2.55)$$

and

$$maxf(p_1, p_2) = f(b_1, a_2). \qquad (2.56)$$

Now assume that $1 - a_1$ does not equal b_2 or $1 - b_1$ is not equal to a_2. Then the optimization problem is not so simple and we will need to use the graphical procedure presented above, graphing the surface $f(p_1, p_2)$ over the rectangle $a_1 \leq p_i \leq b_i$ for $i = 1, 2$, to approximate $max/minf(p_1, p_2)$.

Example 2.9.1.2

Now we have the problem

$$max/minf(p_1, p_2, p_3), \qquad (2.57)$$

subject to

$$p_i \in [a_i, b_i], 1 \leq i \leq 3, p_1 + p_2 + p_3 = 1, \qquad (2.58)$$

where $0 \le a_i \le b_i \le 1$, $i = 1, 2, 3$. Also assume that $\partial f/\partial p_1 > 0$, $\partial f/\partial p_2 < 0$ and $\partial f/\partial p_3 < 0$. If $a_1 + b_2 + b_3 = 1$ and $b_1 + a_2 + a_3 = 1$, then the solution is

$$min f(p_1, p_2, p_3) = f(a_1, b_2, b_3), \tag{2.59}$$

and

$$max f(p_1, p_2, p_3) = f(b_1, a_2, a_3). \tag{2.60}$$

When these sums do not add up to one, we need to employ some numerical optimization method.

Example 2.9.1.3

The last problem is

$$max/min f(p_1, p_3), \tag{2.61}$$

subject to

$$p_i \in [a_i, b_i], 1 \le i \le 3, p_1 + p_2 + p_3 = 1, \tag{2.62}$$

where $0 \le a_i \le b_i \le 1$, $1 \le i \le 3$. Assume that $\partial f/\partial p_1 > 0$ and $\partial f/\partial p_3 < 0$. Also assume that: $(1) p_1 = a_1$ and $p_3 = b_3$ are feasible, or $a_1 + p_2 + b_3 = 1$ for some $p_2 \in [a_2, b_2]$; and (2) $p_1 = b_1$ and $p_3 = a_3$ are feasible, or $b_1 + p_2 + a_3 = 1$ for some $p_2 \in [a_2, b_2]$. Then the solution is

$$min f(p_1, p_3) = f(a_1, b_3), \tag{2.63}$$

and

$$max f(p_1, p_3) = f(b_1, a_3). \tag{2.64}$$

We first try the calculus method and if that procedure is not going to work, then we go to the graphical method. The graphical, or calculus, method was applicable to all of these types of problems in this book. However, one can easily consider situations where neither procedure is applicable. Then you will need a numerical optimization algorithm for non-linear functions having both inequality and equality constraints.

2.9.2 Second Problem

This type of problem is

$$max/min f(\theta), \tag{2.65}$$

subject to

$$\theta_i \in [a_i, b_i], 1 \le i \le n, \tag{2.66}$$

where $\theta = (\theta_1, ..., \theta_n)$. Notice in this case we do not have the constraint that the θ_i must sum to one. In applications of this problem the intervals $[a_i, b_i]$ are α-cuts of fuzzy numbers used for uncertain parameters in probability density (mass) functions. This problem is actually solving to obtain the α-cuts of a fuzzy probability.

In this subsection we explain how we will obtain numerical solutions to this problem. In this book n will be one or two, and n will be two only for the normal probability density function. When $n = 1$ we may employ a calculus method (Example 2.9.2.2 below) or a graphical procedure (discussed in the previous subsection). When $n = 2$ we used the graphical method (see Example 8.3.1). To see more detail on this type of problem let us look at the next two examples.

Example 2.9.2.1

Let $N(\mu, \sigma^2)$ be the normal probability density with mean μ and variance σ^2. To obtain the fuzzy normal we use fuzzy numbers $\overline{\mu}$ and $\overline{\sigma}^2$ for μ and σ^2, respectively. Set $\overline{P}[c, d]$ to be the fuzzy probability of obtaining a value in the interval $[c, d]$. Its α-cuts are gotten by solving the following problem (see Section 8.3)

$$max/min f(\mu, \sigma^2) = \int_{z_1}^{z_2} N(0, 1)dx, \qquad (2.67)$$

subject to

$$\mu \in [a_1, b_1], \sigma^2 \in [a_2, b_2], \qquad (2.68)$$

where $a_i \leq b_i$, $1 \leq i \leq 2$, $a_2 > 0$, and $z_1 = (d - \mu)/\sigma$ and $z_2 = (c - \mu)/\sigma$, and $N(0, 1)$ is the normal with zero mean and unit variance. We use the graphical method, discussed above, to solve this problem.

Example 2.9.2.2

The negative exponential has density $f(x; \lambda) = \lambda \exp(-\lambda x)$ for $x \geq 0$, and the density is zero for $x < 0$. The fuzzy negative exponential has a fuzzy number, say $\overline{\lambda} = (2/4/6)$, substituted for crisp λ. We wish to calculate the fuzzy probability of obtaining a value in the interval $[6, 10]$. Let this fuzzy probability be $\overline{P}[6, 10]$ and its α-cuts , see Section 8.4, are determined from the following problem

$$max/min f(\lambda) = \int_6^{10} \lambda \exp(-\lambda x)dx, \qquad (2.69)$$

subject to

$$\lambda \in [a, b], \qquad (2.70)$$

where $[a, b]$ will be an α-cut of $(2/4/6)$. This problem is easy to solve because $f(\lambda)$ is a decreasing function of λ, $df/d\lambda < 0$, across the interval $[a, b]$ (which is a subset of $[2,6]$). Hence,

$$minf(\lambda) = f(b), \qquad (2.71)$$

and

$$maxf(\lambda) = f(a). \qquad (2.72)$$

2.10   Figures

Some of the figures, graphs of fuzzy probabilities, in the book are difficult to obtain so they were created using different methods. The graphs, except those discussed below, were done first in Maple [18] and then exported to $LaTeX2_\epsilon$. We did many figures first in Maple because of the "implicitplot" command in Maple. Let us explain why this command was so important in Maple. Suppose $\overline{P}$ is a fuzzy probability we want to graph. Throughout this book we determine $\overline{P}$ by first calculating its α-cuts. Let $\overline{P}[\alpha] = [p_1(\alpha), p_2(\alpha)]$. So we get $x = p_1(\alpha)$ describing the left side of the triangular shaped fuzzy number $\overline{P}$ and $x = p_2(\alpha)$ describes the right side. On a graph we would have the x-axis horizontal and the y-axis vertical. α is on the y-axis between zero and one. Substituting y for α we need to graph $x = p_i(y)$, for $i = 1, 2$. But this is backwards, we usually have y a function of x. The "implicitplot" command allows us to do the correct graph with x a function of y.

The following figures were done using the graphics package in $LaTeX2_\epsilon$: Figure 2.5, Figure 2.6, Figure 3.1, Figure 4.2, Figure 4.5, Figure 8.1, Figure 8.2, and Figure 8.5. For certain technical reasons certain Figures (3.1, 4.2, 4.5, and 8.1) could not be completed using only Maple and so they were copied over, and completed, in $LaTeX2_\epsilon$. For the two Figures 8.2 and 8.5 Maple was used to compute the α-cuts and then the graphs were done in $LaTeX2_\epsilon$.

2.11   References

1. G.Bortolon and R.Degani: A Review of Some Methods for Ranking Fuzzy Subsets, Fuzzy Sets and Systems, 15(1985), pp. 1-19.

2. J.J.Buckley: Ranking Alternatives Using Fuzzy Numbers, Fuzzy Sets and Systems, 15(1985), pp.21-31.

3. J.J.Buckley: Fuzzy Hierarchical Analysis, Fuzzy Sets and Systems, 17(1985), pp. 233-247.

4. J.J.Buckley and E.Eslami: Introduction to Fuzzy Logic and Fuzzy Sets, Physica-Verlag, Heidelberg, Germany, 2002.

5. J.J.Buckley and T.Feuring: Fuzzy and Neural: Interactions and Applications, Physica-Verlag, Heidelberg, Germany, 1999.

6. J.J.Buckley and T.Feuring: Evolutionary Algorithm Solutions to Fuzzy Problems: Fuzzy Linear Programming, Fuzzy Sets and Systems, 109(2000), pp. 35-53.

7. J.J. Buckley and Y. Hayashi: Can Neural Nets be Universal Approximators for Fuzzy Functions? Fuzzy Sets and Systems, 101 (1999), pp. 323-330.

8. J.J. Buckley and Y. Qu: On Using α–cuts to Evaluate Fuzzy Equations, Fuzzy Sets and Systems, 38 (1990), pp. 309-312.

9. J.J.Buckley, E.Eslami and T.Feuring: Fuzzy Mathematics in Economics and Engineering, Physica-Verlag, Heidelberg, Germany, 2002.

10. J.J.Buckley, T.Feuring and Y.Hayashi: Solving Fuzzy Problems in Operations Research, J. Advanced Computational Intelligence, 3(1999), pp. 171-176.

11. J.J.Buckley, T.Feuring and Y.Hayashi: Multi-Objective Fully Fuzzified Linear Programming, Int. J. Uncertainty, Fuzziness and Knowledge Based Systems, 9(2001), pp. 605-622

12. J.J.Buckley, T.Feuring and Y.Hayashi: Fuzzy Queuing Theory Revisited", Int. J. Uncertainty, Fuzziness and Knowledge Based Systems, 9(2001), pp. 527-538.

13. J.J.Buckley, T.Feuring and Y.Hayashi: Solving Fuzzy Problems in Operations Research: Inventory Control, Soft Computing. To appear.

14. P.T.Chang and E.S.Lee: Fuzzy Arithmetic and Comparison of Fuzzy Numbers, in: M.Delgado, J,Kacprzyk, J.L.Verdegay and M.A.Vila (eds.), Fuzzy Optimization: Recent Advances, Physica-Verlag, Heidelberg, Germany, 1994, pp. 69-81.

15. D.Dubois, E.Kerre, R.Mesiar and H.Prade: Fuzzy Interval Analysis, in: D.Dubois and H.Prade (eds.), Fundamentals of Fuzzy Sets, The Handbook of Fuzzy Sets, Kluwer Acad. Publ., 2000, pp. 483-581.

16. A.M.Geoffrion: Proper Efficiency and the Theory of Vector Maximization, J. Math. Analysis and Appl., 22(1968), pp. 618-630.

17. G.J. Klir and B. Yuan: Fuzzy Sets and Fuzzy Logic: Theory and Applications, Prentice Hall, Upper Saddle River, N.J., 1995.

18. Maple 6, Waterloo Maple Inc., Waterloo, Canada.

19. R.E. Moore: Methods and Applications of Interval Analysis, SIAM Studies in Applied Mathematics, Philadelphia, 1979.

20. A. Neumaier: Interval Methods for Systems of Equations, Cambridge University Press, Cambridge, U.K., 1990.

21. H.A.Taha: Operations Research, Fifth Edition, Macmillan, N.Y., 1992.

22. X.Wang and E.E.Kerre: Reasonable Properties for the Ordering of Fuzzy Quantities (I), Fuzzy Sets and Systems, 118(2001), pp. 375-385.

23. X.Wang and E.E.Kerre: Reasonable Properties for the Ordering of Fuzzy Quantities (II), Fuzzy Sets and Systems, 118(2001), pp. 387-405.

Chapter 3

Fuzzy Probability Theory

3.1 Introduction

This chapter is based on [1]. However, many new results have been added mostly on fuzzy conditional probability (Section 3.3), fuzzy independence (Section 3.4), fuzzy Bayes' formula (Section 3.5) and the applications in Section 3.6.

Let $X = \{x_1, ..., x_n\}$ be a finite set and let P be a probability function defined on all subsets of X with $P(\{x_i\}) = a_i$, $1 \leq i \leq n$, $0 < a_i < 1$ all i and $\sum_{i=1}^{n} a_i = 1$. X together with P is a discrete (finite) probability distribution. In practice all the a_i values must be known exactly. Many times these values are estimated, or they are provided by experts. We now assume that some of these a_i values are uncertain and we will model this uncertainty using fuzzy numbers. Not all the a_i need to be uncertain, some may be known exactly and are given as a crisp (real) number. If an a_i is crisp, then we will still write it as a fuzzy number even though this fuzzy number is crisp.

Due to the uncertainty in the a_i values we substitute $\bar{a}_i$, a fuzzy number, for each a_i and assume that $0 < \bar{a}_i < 1$ all i. Throughout the rest of this book, if some probability has been estimated from data or from experts, we will use a fuzzy number for this probability. If some a_i is known precisely, then this $\bar{a}_i = a_i$ but we still write a_i as $\bar{a}_i$. Then X together with the $\bar{a}_i$ values is a discrete (finite) fuzzy probability distribution. We write $\overline{P}$ for fuzzy P and we have $\overline{P}(\{x_i\}) = \bar{a}_i$, $1 \leq i \leq n$, $0 < \bar{a}_i < 1$, $1 \leq i \leq n$.

The uncertainty is in some of the a_i values but we know that we have a discrete probability distribution. So we now put the following restriction on the $\bar{a}_i$ values: there are $a_i \in \bar{a}_i[1]$ so that $\sum_{i=1}^{n} a_i = 1$. That is, we can choose a_i in $\bar{a}_i[\alpha]$, all α, so that we get a discrete probability distribution.

3.2   Fuzzy Probability

Let A and B be (crisp) subsets of X. We know how to compute $P(A)$ and $P(B)$ so let us find $\overline{P}(A)$ and $\overline{P}(B)$. To do this we introduce restricted fuzzy arithmetic. There may be uncertainty in some of the a_i values, but there is no uncertainty in the fact that we have a discrete probability distribution. That is, whatever the a_i values in $\overline{a}_i[\alpha]$ we must have $a_1 + ... + a_n = 1$. This is the basis of our restricted fuzzy arithmetic. Suppose $A = \{x_1, ..., x_k\}$, $1 \leq k < n$, then define

$$\overline{P}(A)[\alpha] = \{\sum_{i=1}^{k} a_i | \ \mathbf{S} \ \}, \tag{3.1}$$

for $0 \leq \alpha \leq 1$, where $\mathbf{S}$ stands for the statement "$a_i \in \overline{a}_i[\alpha]$, $1 \leq i \leq n$, $\sum_{i=1}^{n} a_i = 1$ ". This is our restricted fuzzy arithmetic. Notice that we first choose a complete discrete probability distribution from the α-cuts before we compute a probability in equation (3.1). Notice also that $\overline{P}(A)[\alpha]$ is not the sum of the intervals $\overline{a}_i[\alpha]$, $1 \leq i \leq k$ using interval arithmetic. We now show that the $\overline{P}(A)[\alpha]$ are the α-cuts of a fuzzy number $\overline{P}(A)$. But first we require some definitions.

Define

$$S = \{(x_1, ..., x_n)|x_i \geq 0 \ all \ i, \sum_{i=1}^{n} x_i = 1\}, \tag{3.2}$$

and then also define

$$Dom[\alpha] = (\prod_{i=1}^{n} \overline{a}_i[\alpha]) \bigcap S, \tag{3.3}$$

for $0 \leq \alpha \leq 1$. In equation (3.3) we first take the product of n closed intervals producing a "rectangle" in n dimensional space which is then intersected with the set S. Now define a function f mapping $Dom[\alpha]$ into the real numbers as

$$f(a_1, ..., a_n) = \sum_{i=1}^{k} a_i, \tag{3.4}$$

for $(a_1, ..., a_n) \in Dom[\alpha]$. f is continuous, $Dom[\alpha]$ is connected, closed, and bounded implying that the range of f is a closed, and bounded, interval of real numbers. Define

$$\Gamma[\alpha] = f(Dom[\alpha]), \tag{3.5}$$

for $0 \leq \alpha \leq 1$. But, from equation (3.1), we see that

$$\overline{P}(A)[\alpha] = \Gamma[\alpha], \tag{3.6}$$

for all α. Hence, $\overline{P}(A)$ is a fuzzy number since it is normalized ($\overline{P}(A)[1] \neq \phi$).

We can now argue that:

1. If $A \cap B = \phi$, then $\overline{P}(A) + \overline{P}(B) \geq \overline{P}(A \cup B)$.

2. If $A \subseteq B$ and $\overline{P}(A)[\alpha] = [p_{a1}(\alpha), p_{a2}(\alpha)]$ and $\overline{P}(B)[\alpha] = [p_{b1}(\alpha), p_{b2}(\alpha)]$, then $p_{ai}(\alpha) \leq p_{bi}(\alpha)$ for $i = 1, 2$ and $0 \leq \alpha \leq 1$.

3. $0 \leq \overline{P}(A) \leq 1$ all A with $\overline{P}(\phi) = 0$, $\overline{P}(X) = 1$, crisp "one".

4. $\overline{P}(A) + \overline{P}(A') \geq 1$, crisp "one", where A' is the complement of A.

5. When $A \cap B \neq \phi$, $\overline{P}(A \cup B) \leq \overline{P}(A) + \overline{P}(B) - \overline{P}(A \cap B)$.

It is easy to see that (2) and (3) are true and (4) follows from (1) and (3). So we now demonstrate that (1) and the generalized addition law (5) are true. Then we show by Example 3.2.1 below that in cases (1) and (5) we may not get equality.

We show that $(\overline{P}(A) + \overline{P}(B))[\alpha] = \overline{P}(A)[\alpha] + \overline{P}(B)[\alpha] \supseteq \overline{P}(A \cup B)[\alpha]$, for all α. To simplify the discussion assume that $A = \{x_1, ..., x_k\}$, $B = \{x_l, ..., x_m\}$ for $1 \leq k < l \leq m \leq n$. Again let **S** denote the statement "$a_i \in \bar{a}_i[\alpha]$, $1 \leq i \leq n$, $\sum_{i=1}^{n} a_i = 1$". Then we need to show, based on equation (3.1), that

$$\{\sum_{i=1}^{k} a_i | \text{ } \textbf{S} \text{ } \} + \{\sum_{i=l}^{m} a_i | \text{ } \textbf{S} \text{ } \} \supseteq \{\sum_{i=1}^{k} a_i + \sum_{i=l}^{m} a_i | \text{ } \textbf{S} \text{ } \}. \qquad (3.7)$$

Let $r = s + t$ be a member of the right side of equation (3.7) where $s = a_1 + ... + a_k$ and $t = a_l + ... + a_m$. Then s belongs to the first member of the left side of equation (3.7) and t belongs to the second member. Hence $r = s + t$ belongs to the left side of equation (3.7) and equation (3.7) is correct.

However, there are situations where $\overline{P}(A) + \overline{P}(B) = \overline{P}(A \cup B)$ when A and B are disjoint. We also give an example of equality in Example 3.2.1 below.

Next we wish to show (5) is also true. Using the notation defined above assume that $A = \{x_1, ..., x_k\}$, $B = \{x_l, ..., x_m\}$ but now $1 \leq l \leq k \leq m \leq n$. We show that $\overline{P}(A)[\alpha] + \overline{P}(B)[\alpha] - \overline{P}(A \cap B)[\alpha] \supseteq \overline{P}(A \cup B)[\alpha]$. Or, we show that

$$\{\sum_{i=1}^{k} a_i | \textbf{S}\} + \{\sum_{i=l}^{m} a_i | \textbf{S}\} - \{\sum_{i=l}^{k} a_i | \textbf{S}\} \supseteq \{\sum_{i=1}^{m} a_i | \textbf{S}\}. \qquad (3.8)$$

Let r be in the right side of equation (3.8). Then we may write r as $r = s + t - u$ where $s = a_1 + ... + a_k$, $t = a_l + ... + a_m$ and $u = a_l + ... + a_k$. Now s belongs to the first member on the left side of equation (3.8), t belongs to the second member and u belongs to the third member. Hence r belongs to the left side of equation (3.8).

Example 3.2.1

We first show by example that you may not obtain equality in equation (3.7). Let $n = 5$, $A = \{x_1, x_2\}$, $B = \{x_4, x_5\}$, $a_i = 0.2$ for $1 \le i \le 5$. All the probabilities are uncertain except a_3 so let $\bar{a}_1 = \bar{a}_2 = (0.19/0.2/0.21)$, $\bar{a}_3 = 0.2$ and $\bar{a}_4 = \bar{a}_5 = (0.19/0.2/0.21)$. Then $\overline{P}(A)[0] = [0.38, 0.42]$ because $p_1 = 0.19 = p_2$ are feasible (see Section 2.9)and $p_1 = p_2 = 0.21$ are feasible. We also determine $\overline{P}(B)[0] = [0.38, 0.42]$ so the left side of equation (3.7), for $\alpha = 0$, is the interval $[0.76, 0.84]$. However, $\overline{P}(A \cup B)[0] = [0.8, 0.8]$. For A and B disjoint, we can get $\overline{P}(A \cup B)[\alpha]$ a proper subset of $\overline{P}(A)[\alpha] + \overline{P}(B)[\alpha]$.

Let $n = 6$, $A = \{x_1, x_2, x_3\}$, $B = \{x_3, x_4, x_5\}$, $a_i = 0.1$ for $1 \le i \le 5$ and $a_6 = 0.5$, Assuming all probabilities are uncertain we substitute $\bar{a}_i = (0.05/0.1/0.15)$ for $1 \le i \le 5$ and $\bar{a}_6 = (0.25/0.5/0.75)$. Then we easily deduce that $\overline{P}(A \cup B)[0] = [0.25, 0.75]$, $\overline{P}(A)[0] = \overline{P}(B)[0] = [0.15, 0.45]$ and $\overline{P}(A \cap B)[0] = [0.05, 0.15]$. Then, from interval arithmetic, we see that

$$[0.25, 0.75] \ne [0.15, 0.45] + [0.15, 0.45] - [0.05, 0.15], \qquad (3.9)$$

where the right side of this equation is the interval $[0.15, 0.85]$. So $\overline{P}(A \cup B)[\alpha]$ can be a proper subset of $\overline{P}(A)[\alpha] + \overline{P}(B)[\alpha] - \overline{P}(A \cap B)[\alpha]$.

Now we show by example we can obtain $\overline{P}(A) + \overline{P}(B) = \overline{P}(A \cup B)$ when A and B are disjoint. Let $X = \{x_1, x_2, x_3\}$, $A = \{x_1\}$, $B = \{x_3\}$, $\bar{a}_1 = (0.3/0.33/0.36)$, $\bar{a}_2 = (0.28/0.34/0.40)$ and $\bar{a}_3 = \bar{a}_1$. Then $\overline{P}(A) = \bar{a}_1$, $\overline{P}(B) = \bar{a}_3$ so $\overline{P}(A) + \overline{P}(B) = (0.6/0.66/0.72)$. Alpha-cuts of $\overline{P}(A \cup B)$ are

$$\overline{P}(A \cup B)[\alpha] = \{a_1 + a_3| \quad \textbf{S} \quad \}. \qquad (3.10)$$

Let $\bar{a}_i[\alpha] = [a_{i1}(\alpha), a_{i2}(\alpha)]$, for $i = 1, 2, 3$. We can evaluate equation (3.10) by using the end points of these α-cuts because; (1) for any α there is a $a_2 \in \bar{a}_2[\alpha]$ so that $a_{11}(\alpha) + a_2 + a_{31}(\alpha) = 1$; and (2) for each α there is a $a_2 \in \bar{a}_2[\alpha]$ so that $a_{12}(\alpha) + a_2 + a_{32}(\alpha) = 1$. Then

$$\overline{P}(A \cup B)[\alpha] = [a_{11}(\alpha) + a_{31}(\alpha), a_{12}(\alpha) + a_{32}(\alpha)], \qquad (3.11)$$

so that $\overline{P}(A \cup B) = (0.6/0.66/0/72)$.

We will finish this section with the calculation of the mean and variance of a discrete fuzzy probability distribution. The fuzzy mean is defined by its α-cuts

$$\bar{\mu}[\alpha] = \{\sum_{i=1}^{n} x_i a_i| \quad \textbf{S} \quad \}, \qquad (3.12)$$

where, as before, $\textbf{S}$ denotes the statement " $a_i \in \bar{a}_i[\alpha], 1 \le i \le n, \sum_{i=1}^{n} a_i = 1$". The variance is also defined by its α-cuts as

$$\bar{\sigma}^2[\alpha] = \{\sum_{i=1}^{n} (x_i - \mu)^2 a_i| \quad \textbf{S} \,, \mu = \sum_{i=1}^{n} x_i a_i\}. \qquad (3.13)$$

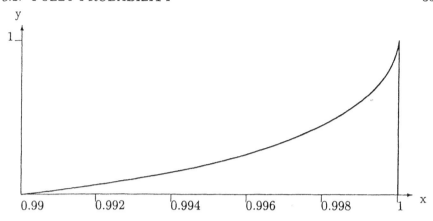

Figure 3.1: Fuzzy Variance in Example 3.2.2

The mean $\bar{\mu}$ and variance $\bar{\sigma}^2$ will be fuzzy numbers because $\bar{\mu}[\alpha]$ and $\bar{\sigma}^2[\alpha]$ are closed, bounded, intervals for $0 \leq \alpha \leq 1$. The same proof, as was given for $\overline{P}(A)[\alpha]$, can be used to justify these statements.

Example 3.2.2

Let $X = \{0,1,2,3,4\}$ with $a_0 = a_4 = \frac{1}{16}$, $a_1 = a_3 = 0.25$ and $a_2 = \frac{3}{8}$. Then $\mu = 2$ and $\sigma^2 = 1$. Assume there is uncertainty only in a_1 and a_3 so we substitute $\bar{a}_1$ for a_1 and $\bar{a}_3$ for a_3. Let us use $\bar{a}_1 = \bar{a}_3 = (0.2/0.25/0.3)$. First compute $\bar{\mu}[\alpha]$. Use the numerical values for the x_i, a_0, a_2 and a_4 but choose $a_1 \in \bar{a}_1[\alpha]$ and $a_3 = 0.5 - a_1$ in $\bar{a}_3[\alpha]$ so that the sum of the a_i equals one. Then the formula for crisp $\mu = f_1(a_1) = 2.5 - 2a_1$ is a function of only a_1. We see that $\partial f_1/\partial a_1 < 0$. This allows us to compute the end points of the interval $\bar{\mu}[\alpha]$ which gives $[1.9 + 0.1\alpha, 2.1 - 0.1\alpha] = \bar{\mu}[\alpha]$, so that $\bar{\mu} = (1.9/2/2.1)$, a triangular fuzzy number. Since $\bar{a}_1[\alpha] = [0.2 + 0.05\alpha, 0.3 - 0.05\alpha]$, we used $0.3 - 0.05\alpha$ to get $1.9 + 0.1\alpha$ and $0.2 + 0.05\alpha$ to obtain $2.1 - 0.1\alpha$. We do the same with the crisp formula for σ^2 and we deduce that $\sigma^2 = f_2(a_1) = 0.75 + 2a_1 - 4a_1^2$, for a_1 in $\bar{a}_1[\alpha]$. If $\bar{\sigma}^2[\alpha] = [\sigma_1^2(\alpha), \sigma_2^2(\alpha)]$ we determine from $f_2(a_1)$ that $\sigma_1^2(\alpha) = f_2(0.2 + 0.05\alpha)$ but $\sigma_2^2(\alpha) = 1$ all α. So the α-cuts of the fuzzy variance are $[0.99 + 0.02\alpha - 0.01\alpha^2, 1]$, $0 \leq \alpha \leq 1$. The uncertainty in the variance is that it can be less than one but not more than one. The graph of the fuzzy variance is shown in Figure 3.1.

It can be computationally difficult, in general, to compute the intervals $\overline{P}(A)[\alpha]$ (equation (3.1)), $\bar{\mu}[\alpha]$ (equation (3.12)), and $\bar{\sigma}^2[\alpha]$ (equation (3.13)). All we need to do is to determine the end points of these intervals which can be written as a non-linear optimization problem. For example, for $\bar{\sigma}^2[\alpha]$, we have

$$\sigma_1^2(\alpha) = min\{\sum_{i=1}^{n}(x_i - \mu)^2 a_i| \quad \mathbf{S} \quad \},\tag{3.14}$$

and

$$\sigma_2^2(\alpha) = max\{\sum_{i=1}^{n}(x_i - \mu)^2 a_i| \quad \mathbf{S} \quad \},\qquad(3.15)$$

where $\mathbf{S}$ is the statement "$a_i \in \bar{a}_i[\alpha], 1 \le i \le n, \mu = \sum_{i=1}^{n} x_i a_i, \sum_{i=1}^{n} a_i = 1$". One may consider a directed search algorithm (genetic, evolutionary) to estimate the $\sigma_i^2(\alpha)$ values for selected α, or see Section 2.9.

3.3   Fuzzy Conditional Probability

Let $A = \{x_1, ..., x_k\}$, $B = \{x_l, ..., x_m\}$ for $1 \le l \le k \le m \le n$ so that A and B are not disjoint. We wish to define the fuzzy conditional probability of A given B. We will write this fuzzy conditional probability as $\overline{P}(A|B)$. We now present two definitions for fuzzy conditional probability and then argue in favor of the first definition.

Our first definition is

$$\overline{P}(A|B) = \{\frac{\sum_{i=l}^{k} a_i}{\sum_{i=l}^{m} a_i}| \quad \mathbf{S} \quad \}.\qquad(3.16)$$

In this definition the numerator of the quotient is the sum of the a_i in the intersection of A and B, while the denominator is the sum of the a_i in B.

Our second definition is

$$\overline{P}(A|B) = \frac{\overline{P}(A \cap B)}{\overline{P}(B)}.\qquad(3.17)$$

This second definition seems very natural but, as to be shown in Example 3.3.1 below, because of fuzzy arithmetic this conditional fuzzy probability can get outside the interval $[0, 1]$. The first definition always produces a fuzzy probability in $[0, 1]$.

Example 3.3.1

Let $n = 4$, $A = \{x_1, x_2\}$, $B = \{x_2, x_3\}$ and all the a_i are uncertain with $\bar{a}_1 = (0.1/0.2/0.3)$, $\bar{a}_2 = (0.2/0.3/0.4)$, $\bar{a}_3 = (0/0.1/0.2)$ and $\bar{a}_4 = (0.3/0.4/0.5)$. We show that, using the second definition, $\overline{P}(A|B)$ is not entirely in $[0, 1]$

Since $A \cap B = \{x_2\}$ we find that $\overline{P}(A \cap B) = \bar{a}_2$. We also easily find that $\overline{P}(B) = (0.2/0.4/0.6)$. Then this definition produces a fuzzy number $[(0.2/0.3/0.4)]/[(0.2/0.4/0.6)]$ whose $\alpha = 0$ cut is the interval $[\frac{1}{3}, 2]$ with right end point greater than one. We would expect this to occur quite often using the second definition so we drop the second definition and adopt the first definition for fuzzy conditional probability.

Example 3.3.2

We will use the same data as given in Example 3.3.1 in the first definition to compute the fuzzy conditional probability of A given B. We need to evaluate

$$\overline{P}(A|B) = \{\frac{a_2}{a_2 + a_3}|\ \mathbf{S}\ \}, \tag{3.18}$$

for $0 \leq \alpha \leq 1$. If we let $y = f(a_2, a_3) = \frac{a_2}{a_2+a_3}$ we find that $\partial y / \partial a_2 > 0$ and $\partial y / \partial a_3 < 0$. This allows us to find the end points of the interval defining the α-cuts of the fuzzy conditional probability. We obtain

$$\overline{P}(A|B)[\alpha] = [0.5 + 0.25\alpha, 1 - 0.25\alpha], \tag{3.19}$$

for all $\alpha \in [0, 1]$. Let $\overline{a}_i[\alpha] = [a_{i1}(\alpha), a_{i2}(\alpha)]$ for $1 \leq i \leq 4$. We see that; (1) $a_{21}(\alpha)$ and $a_{32}(\alpha)$ are feasible; and (2) $a_{22}(\alpha)$ and $a_{31}(\alpha)$ are feasible. What we did, to obtain the left end point of the interval in equation (3.19), was in the function f we : (1) substituted the left end point of the interval $\overline{a}_2[\alpha]$, which is $0.2 + 0.1\alpha$, for a_2; and (2) substituted the right end point of the interval $\overline{a}_3[\alpha]$, which is $0.2 - 0.1\alpha$, for a_3. To get the right end point of the interval: (1) substitute the right end point of the interval $\overline{a}_2[\alpha]$ for a_2; and (2) substitute the left end point of $\overline{a}_3[\alpha]$ for a_3. Hence, $\overline{P}(A|B) = (0.5/0.75/1)$ a triangular fuzzy number.

We will use the first definition of fuzzy conditional probability in the remainder of this book. Now we will show the basic properties of fuzzy conditional probability which are:

1. $0 \leq \overline{P}(A|B) \leq 1$;

2. $\overline{P}(B|B) = 1$, crisp one;

3. $\overline{P}(A_1 \cup A_2|B) \leq \overline{P}(A_1|B) + \overline{P}(A_2|B)$, if $A_1 \cap A_2 = \phi$;

4. $\overline{P}(A|B) = 1$, crisp one, if $B \subseteq A$; and

5. $\overline{P}(A|B) = 0$, crisp zero, if $B \cap A = \phi$.

Items (1), (2) and (4) follow immediately from our definition of fuzzy conditional probability. If we define the value of an empty sum to be zero, then (5) is true since the numerator in equation (3.16) will be empty when A and B are disjoint. So let us only present an argument for property (3). We show that $\overline{P}(A_1 \cup A_2|B)[\alpha] \subseteq \overline{P}(A_1|B)[\alpha] + \overline{P}(A_2|B)[\alpha]$, for $0 \leq \alpha \leq 1$. Choose and fix α. Let x belong to $\overline{P}(A_1 \cup A_2|B)[\alpha]$. Then $x = \frac{r+s}{t}$ where: (1) r is the sum of the a_i for $x_i \in A_1 \cap B$; (2) s is the sum of the a_i for $x_i \in A_2 \cap B$; and (3) t is the sum of the a_i for $x_i \in B$. Of course, the $a_i \in \overline{a}_i[\alpha]$ all i and the sum of the a_i equals one. Then $r/t \in \overline{P}(A_1|B)[\alpha]$ and $s/t \in \overline{P}(A_2|B)[\alpha]$. Hence $x \in \overline{P}(A_1|B)[\alpha] + \overline{P}(A_2|B)[\alpha]$ and the result follows. The following example shows that we may obtain $\overline{P}(A_1 \cup A_2|B) \neq \overline{P}(A_1|B) + \overline{P}(A_2|B)$ when $A_1 \cap A_2 = \phi$.

Example 3.3.3

We will use the same data as in Examples 3.3.1 and 3.3.2 but let $A_1 = \{x_2\}$ and $A_2 = \{x_3\}$ so that $A_1 \cup A_2 = B$ and $\overline{P}(A_1 \cup A_2|B) = 1$. We quickly determine, as in Example 3.3.2 that $\overline{P}(A_1|B) = (0.5/0.75/1)$ and $\overline{P}(A_2|B) = (0/0.25/0.5)$ so that $\overline{P}(A_1|B) + \overline{P}(A_2|B) = (0.5/1/1.5)$. Notice that when we add two fuzzy numbers both in $[0, 1]$ we can get a fuzzy number not contained in $[0, 1]$. Clearly, $\overline{P}(A_1 \cup A_2|B) \leq \overline{P}(A_1|B) + \overline{P}(A_2|B)$ but they are not equal.

3.4   Fuzzy Independence

We will present two definitions of two events A and B being independent and two versions of the first definition. We then argue against the second definition and therefore adopt the first definition for this book.

The first definition uses fuzzy conditional probability from the previous section. We will say that A and B are strongly independent if

$$\overline{P}(A|B) = \overline{P}(A), \tag{3.20}$$

and

$$\overline{P}(B|A) = \overline{P}(B). \tag{3.21}$$

Since the equality in equations (3.20) and (3.21) are sometimes difficult to satisfy we also have a weaker definition of independence. We will say that A and B are weakly independent if

$$\overline{P}(A|B)[1] = \overline{P}(A)[1], \tag{3.22}$$

and

$$\overline{P}(B|A)[1] = \overline{P}(B)[1]. \tag{3.23}$$

In the weaker version of independence we only require the equality for the core (where the membership values are one) of the fuzzy numbers. Clearly, if they are strongly independent they are weakly independent.

Our second definition of independence follows from the usual way of specifying independence in the crisp (non-fuzzy) case. We say that A and B are independent if

$$\overline{P}(A \cap B) = \overline{P}(A)\overline{P}(B). \tag{3.24}$$

We now argue in favor of the first definition because, due to fuzzy multiplication, it would be very rare to have the equality in equation (3.24) hold. The following example shows that events can be strongly independent (first definition) but it is too difficult for them to be independent by the second definition.

Example 3.4.1

Let $n = 4$ and $\bar{a}_i = (0.2/0.25/0.3)$, $1 \le i \le 4$. Also let $A = \{x_1, x_2\}$ and $B = \{x_2, x_3\}$. First, we easily see that $\overline{P}(A) = (0.4/0.5/0.6) = \overline{P}(B)$. To find $\overline{P}(A|B)$ we need to compute

$$\overline{P}(A|B)[\alpha] = \{\frac{a_2}{a_2 + a_3} \mid \text{ S } \}, \tag{3.25}$$

for all α. We do this as in Example 3.3.2 and obtain $\overline{P}(A|B) = (0.4/0.5/0.6) = \overline{P}(A)$. Similarly we see that $\overline{P}(B|A) = (0.4/0.5/0.6) = \overline{P}(B)$ and A and B are strongly independent. Now we go to the second definition and find $\overline{P}(A \cap B) = \bar{a}_2 = (0.2/0.25/0.3)$. But $\overline{P}(A)\overline{P}(B) \approx (0.16/0.25/0.36)$ a triangular shaped fuzzy number. Even if $\overline{P}(A|B)$, $\overline{P}(A)$ and $\overline{P}(B)$ are all triangular fuzzy numbers, $\overline{P}(A)\overline{P}(B)$ will not be a triangular fuzzy number. Because of fuzzy multiplication we would not expect the second definition of independence to hold, except in rare cases.

We will adopt the first definition of independence. Now let us see what are the basic properties of independence for fuzzy probabilities. In crisp probability theory we know that if A and B are independent so are A and B', and so are A' and B, and so are A' and B', where the "prime" denotes complement. However, this result may or may not be correct for strong independence and fuzzy probabilities. For the data in Example 3.4.1, A and B are strongly independent and so are A and B', as are A' and B, and this is also true for A' and B'. The following example shows that this is not always true.

Example 3.4.2

Let $n = 4$ but now $\bar{a}_i = (0.2/0.25/0.30)$, $1 \le i \le 3$ with $\bar{a}_4 = (0.1/0.25/0.4)$, and $A = \{x_1, x_2\}$, $B = \{x_2, x_3\}$. As in Example 3.4.1 we find that A and B are strongly independent but we can now show that A and B' are not strongly independent. That is, we argue that $\overline{P}(A|B') \ne \overline{P}(A)$. We know that $\overline{P}(A) = (0.4/0.5/0.6)$. To find α-cuts of $\overline{P}(A|B')$ we compute

$$\{\frac{a_1}{a_1 + a_4} \mid \text{ S } \}. \tag{3.26}$$

Now the fraction in equation (3.26) is an increasing function of a_1 but it is decreasing in a_4. Hence we obtain

$$\overline{P}(A|B')[\alpha] = [\frac{0.2 + 0.05\alpha}{0.6 - 0.1\alpha}, \frac{0.3 - 0.05\alpha}{0.4 + 0.1\alpha}], \tag{3.27}$$

for all $\alpha \in [0,1]$. So $\overline{P}(A|B') \approx (\frac{1}{3}/0.5/0.75)$ a triangular shaped fuzzy number. We see that A and B' are not strongly independent.

However, the situation is possibly changed for weakly independent. Suppose all the $\bar{a}_i$ are triangular fuzzy numbers, A and B are not disjoint, $\bar{p}_i[1] = \frac{1}{n}$, $1 \leq i \leq n$, and A and B are weakly independent. Then it is it true that: (1) A and B' are weakly independent; (2) A' and B are weakly independent; and (3) A' and B' are also weakly independent?

3.5   Fuzzy Bayes' Formula

Fuzzy Bayes' formula is also used in Chapter 7. Let A_i, $1 \leq i \leq m$, be a partition of $X = \{x_1, ..., x_n\}$. That is, the A_i are non-empty, mutually disjoint and their union is X. We do not know the probability of the A_i but we do know the conditional probability of A_i given the state of nature. There is a finite set of chance events, also called the states of nature, $S = \{S_1, ..., S_K\}$ over which we have no control. What we do know is

$$a_{ik} = P(A_i|S_k), \tag{3.28}$$

for $1 \leq i \leq m$ and $1 \leq k \leq K$. If the operative state of nature is S_k, then the a_{ik} give the probabilities of the events A_i.

We do not know the probabilities of the states of nature, so we enlist a group of experts to give their estimates of $a_k = P(S_k)$. The a_k, $1 \leq k \leq K$, is called the prior probability distribution over the states of nature. From their estimates we construct fuzzy probabilities $\bar{a}_k$, see Section 2.7. We first present Bayes' formula using the crisp probabilities for the states of nature.

The probability that the state of nature S_k is in force, given the information that outcome A_j has occurred, is given by Bayes' formula

$$P(S_k|A_j) = \frac{P(A_j|S_k)P(S_k)}{\sum_{k=1}^{K} P(A_j|S_k)P(S_k)}, \tag{3.29}$$

for $1 \leq k \leq K$. The $a_{kj} = P(S_k|A_j)$, $1 \leq k \leq K$, is the posterior probability distribution over the states of nature.

Let us see how this result may be used. Using the a_{ik} and the prior distribution a_k, we may calculate $P(A_i)$ as follows

$$P(A_i) = \sum_{k=1}^{K} P(A_i|S_k)P(S_k), \tag{3.30}$$

for $1 \leq i \leq m$. Now we gather some information and observe that event A_j has occurred. We update the prior to the posterior and then obtain improved estimates of the probabilities for the A_i as

$$P(A_i) = \sum_{k=1}^{K} P(A_i|S_k)P(S_k|A_j), \tag{3.31}$$

for $1 \leq i \leq m$.

Now substitute $\bar{a}_k$ for a_k, $1 \leq k \leq K$. Suppose we observe that event A_j has occurred. Alpha-cuts of the fuzzy posterior distribution are

$$\overline{P}(S_k|A_j)[\alpha] = \{ \frac{a_{jk}a_k}{\sum_{k=1}^{K} a_{jk}a_k} \mid \mathbf{S} \}, \qquad (3.32)$$

for $1 \leq k \leq K$, where $\mathbf{S}$ is the statement " $a_k \in \bar{a}_k[\alpha]$, $1 \leq k \leq K$, $\sum_{k=1}^{K} a_k = 1$ ". It may not be difficult to find these α-cuts. Suppose $K = 3$, k=2 and let

$$f(a_1, a_2, a_3) = \frac{a_{j2}a_2}{\sum_{k=1}^{3} a_{jk}a_k}. \qquad (3.33)$$

Then $\partial f/\partial a_1 < 0$, $\partial f/\partial a_2 > 0$ and $\partial f/\partial a_3 < 0$. Let $\bar{a}_k[\alpha] = [a_{k1}(\alpha), a_{k2}(\alpha)]$ for $k = 1, 2, 3$. If $a_{12}(\alpha) + a_{21}(\alpha) + a_{32}(\alpha) = 1$ and $a_{11}(\alpha) + a_{22}(\alpha) + a_{31}(\alpha) = 1$, then we get the left (right) end point of the interval for the α-cut by substituting $a_{12}(\alpha)$, $a_{21}(\alpha)$, $a_{32}(\alpha)$ ($a_{11}(\alpha)$, $a_{22}(\alpha)$, $a_{31}(\alpha)$) for a_1, a_2, a_3, respectively.

Once we have the fuzzy posterior, we may update our fuzzy probability for the A_i.

There is an alternate method of computing fuzzy Bayes' rule. We could just substitute the fuzzy numbers $\overline{P}(S_k) = \bar{a}_k$ into equation (3.29) and compute the result. However, we would come up against the same problem noted in Section 3.3, see Example 3.3.1, where the result can produce a fuzzy number not in the interval $[0, 1]$.

3.6 Applications

We first present two applications of fuzzy probability followed by two applications of fuzzy conditional probability. Then we give an application of fuzzy Bayes' formula. We have a change of notation in this section: we will use p_i for probability values instead of a_i used in the first five sections of this chapter.

3.6.1 Blood Types

There are four basic blood types: A, B, AB and O. A certain city is going to have a blood drive and they want to know that if they select one person at random, from the pool of possible blood donors, what is the probability that this person does not have blood type O? Type O is the universal blood donor group. They conduct a random sample of 1000 people from the pool of blood donors and they determine the following point estimates: (1) $p_a = 0.33$, or 33% have blood type A; (2) $p_b = 0.23$, or 23% have blood type B; (3) $p_{ab} = 0.35$, or 35% are of blood type AB; and (4) $p_o = 0.09$, or 9% belong to blood type O. Because these are point estimates based on a random sample we will substitute fuzzy numbers for these probabilities. Let $\overline{p}_a = (0.3/0.33/0.36)$,

α	$\overline{P}(O')[\alpha]$
0	[0.88,0.94]
0.2	[0.886,0.934]
0.4	[0.892,0.928]
0.6	[0.898,0.922]
0.8	[0.904,0.916]
1.0	0.91

Table 3.1: Alpha-Cuts of $\overline{P}(O')$

$\overline{p}_b = (0.2/0.23/0.26)$, $\overline{p}_{ab} = (0.32/0.35/0.38)$ and $\overline{p}_o = (0.06/0.09/0.12)$. Next let $\overline{P}(O')$ stand for the fuzzy probability of a donor not having blood type O. From the discussion in Section 3.2 we do not expect $\overline{P}(O')$ to equal $1 - \overline{P}(O)$. However, we find α-cuts of this fuzzy probability as

$$\overline{P}(O')[\alpha] = \{p_a + p_b + p_{ab} \mid \textbf{S} \}, \tag{3.34}$$

for all α, where **S** denotes the statement "$p_a \in \overline{p}_a[\alpha]$, $p_b \in \overline{p}_b[\alpha]$, $p_{ab} \in \overline{p}_{ab}[\alpha]$, $p_o \in \overline{p}_o[\alpha]$, $p_a + p_b + p_{ab} + p_o = 1$". Define $\overline{p}_w[\alpha] = [p_{w1}(\alpha), p_{w2}(\alpha)]$ for $w = a, b, ab, o$. Let $\overline{P}(O')[\alpha] = [o_{n1}(\alpha), o_{n2}(\alpha)]$. We would like to substitute $p_{a1}(\alpha) + p_{b1}(\alpha) + p_{ab1}(\alpha)$ for $p_a + p_b + p_{ab}$ in equation (3.34) to get $o_{n1}(\alpha)$, but this set of $p'_i s$ is not feasible. What we mean is that $p_{a1}(0) + p_{b1}(0) + p_{ab1}(0) = 0.82$ and there is no value of $p_0 \in \overline{p}_o[0]$ that can make the sum equal to one. Therefore we need to use some numerical method to compute these α-cuts. We used "simplex" in Maple [2]. The results are displayed in Table 3.1. From the data in Table 3.1 we compute $\overline{P}(O')[\alpha] = [0.88 + 0.03\alpha, 0.94 - 0.03\alpha]$, or it is a triangular fuzzy number $(0.88/0.91/0.94)$, which is shown in Figure 3.2.

3.6.2   Resistance to Surveys

Pollsters are concerned about the increased level of resistance of people to answer questions during surveys. They conduct a study using a random sample of n people, from a large population of people who are candidates for surveys, ages 18 to 40. This random sample is broken up into two subpopulations: (1) a population of n_1 "young" people with ages 18 to 29; and (2) a population n_2 ($n_1 + n_2 = n$) of "older" people ages 30-40. From this data they obtain the following point estimates: (1) $p_1 = 0.18$, or 18% of young people said they will respond to questions in a survey; (1) $p_2 = 0.05$, or 5% of the young people said that they would not respond to questions in a survey; (3) $p_3 = 0.68$, or 68% of the older people responded that they would participate in a survey; and (4) $p_4 = 0.09$, or 9% of the older people indicated that they would not participate in a survey. The pollsters want to know if they choose at random one person from this group, aged 18 to 40, what is the probability

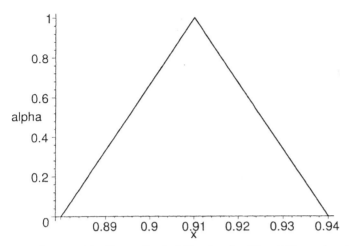

Figure 3.2: Fuzzy Probability in the Blood Type Application

that this person is young or is someone who would refuse to respond to the questions in a survey?

All these probabilities were estimated and so we substitute fuzzy numbers for the p_i. Assume that we decide to use : (1) $\bar{p}_1 = (0.16/0.18/0.20)$ for p_1; (2) $\bar{p}_2 = (0.03/0.05/0.07)$ for p_2; (3) $\bar{p}_3 = (0.61/0.68/0.75)$ for p_3; and (4) $\bar{p}_4 = (0.07/0.09/0.11)$ for p_4. Let A be the event that the person is young and set B to be the event that the person will not respond to a survey. We therefore wish to find the fuzzy probability $\overline{P}(A \cup B)$. From Section 3.2 we know that this fuzzy probability may not equal $\overline{P}(A) + \overline{P}(B) - \overline{P}(A \cap B)$, since A and B are not disjoint. However, we may still find the α-cuts of $\overline{P}(A \cup B)$ as follows

$$\overline{P}(A \cup B)[\alpha] = \{p_1 + p_2 + p_4| \quad \mathbf{S} \quad \},\tag{3.35}$$

for $0 \le \alpha \le 1$, where $\mathbf{S}$ is the statement " $p_i \in \bar{p}_i[\alpha]$, $1 \le i \le 4$, and $p_1 + ... + p_4 = 1$". Let $\bar{p}_i[\alpha] = [p_{i1}(\alpha), p_{i2}(\alpha)]$, $1 \le i \le 4$ and set $\overline{P}(A \cup B)[\alpha] = [P_1(\alpha), P_2(\alpha)]$. Then

$$\overline{P}_1(\alpha) = p_{11}(\alpha) + p_{21}(\alpha) + p_{41}(\alpha),\tag{3.36}$$

and

$$\overline{P}_2(\alpha) = p_{12}(\alpha) + p_{22}(\alpha) + p_{42}(\alpha),\tag{3.37}$$

for all α because now these $p_i's$ are feasible. What this means is: (1) for all α there is a $p_3 \in \bar{p}_3[\alpha]$ so that $p_{11}(\alpha) + p_{21}(\alpha) + p_{41}(\alpha) + p_3 = 1$; and (2) for all α there is a value of $p_3 \in \bar{p}_3[\alpha]$ so that $p_{12}(\alpha) + p_{22}(\alpha) + p_{42}(\alpha) + p_3 = 1$. The graph of the fuzzy probability $\overline{P}(A \cup B)$ is in Figure 3.3. It turns out that $\overline{P}(A \cup B)$ is a triangular fuzzy number $(0.26/0.32/0.38)$.

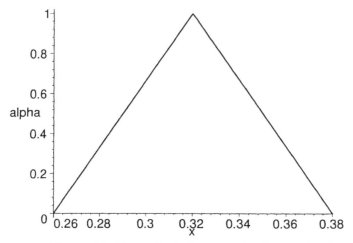

Figure 3.3: Fuzzy Probability in the Survey Application

3.6.3   Testing for HIV

It is important to have an accurate test for the HIV virus. Suppose we have a test, which we shall call T, and we want to see how accurately it predicts that a person has the HIV virus. Let A_1 be the event that a person is infected with the HIV virus, and A_2 is the event that a person is not infected. Also, let B_1 be the event that the test T is "positive" indicating that the person has the virus and set B_2 to be the event that the test T gives the result of "negative", or the person does not have the virus. We want to find the conditional probability of A_1 given B_1, or $P(A_1|B_1)$. To estimate this probability we gather some data. From a large population of the "at-risk" population we take a random sample to estimate the probabilities $p_{11} = P(A_1 \cap B_1)$, $p_{12} = P(A_1 \cap B_2)$, $p_{21} = P(A_2 \cap B_1)$ and $p_{22} = P(A_2 \cap B_2)$. Assume we obtain the estimates $p_{11} = 0.095$, $p_{12} = 0.005$, $p_{21} = 0.045$ and $p_{22} = 0.855$. To show the uncertainty in these point estimates we now substitute fuzzy numbers for the p_{ij}. Let $\bar{p}_{11} = (0.092/0.095/0.098)$, $\bar{p}_{12} = (0.002/0.005/0.008)$, $\bar{p}_{21} = (0.042/0.045/0.048)$ and $\bar{p}_{22} = (0.825/0.855/0.885)$.

The fuzzy probability we want is $\overline{P}(A_1|B_1)$ whose α-cuts are

$$\overline{P}(A_1|B_1)[\alpha] = \{ \frac{p_{11}}{p_{11} + p_{21}} \mid \quad \mathbf{S} \quad \}, \tag{3.38}$$

for all α, where $\mathbf{S}$ is "$p_{ij} \in \bar{p}_{ij}[\alpha]$, $1 \le i, j \le 2$ and $p_{11} + ... + p_{22} = 1$". Let $H(p_{11}, p_{21}) = \frac{p_{11}}{p_{11}+p_{21}}$. We determine that H is an increasing function of p_{11} and it is a decreasing function of p_{21}. Hence, if $\overline{P}(A_1|B_1)[\alpha] = [P_l(\alpha), P_u(\alpha)]$, then

$$P_l(\alpha) = H(p_{111}(\alpha), p_{212}(\alpha)), \tag{3.39}$$

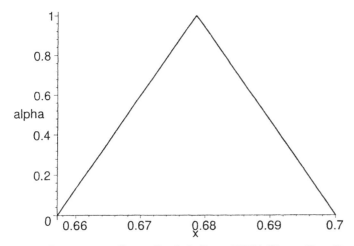

Figure 3.4: Fuzzy Probability of HIV Given Test Positive

and

$$P_u(\alpha) = H(p_{112}(\alpha), p_{211}(\alpha)], \tag{3.40}$$

where $\bar{p}_{ij}[\alpha] = [p_{ij1}(\alpha), p_{ij2}(\alpha)]$, for $1 \le i, j \le 2$. Equations (3.39) and (3.40) are correct because : (1) for any α there are values of $p_{12} \in \bar{p}_{12}[\alpha]$ and $p_{22} \in \bar{p}_{22}[\alpha]$ so that $p_{111}(\alpha) + p_{212}(\alpha) + p_{12} + p_{22} = 1$; and (2) for any α there are values of $p_{12} \in \bar{p}_{12}[\alpha]$ and $p_{22} \in \bar{p}_{22}[\alpha]$ so that $p_{112}(\alpha) + p_{211}(\alpha) + p_{12} + p_{22} = 1$. The graph of $\overline{P}(A_1|B_1)$ is in Figure 3.4. We get a triangular uzzy number $(\frac{92}{140} / \frac{95}{140} / \frac{98}{140})$ for $\overline{P}(A_1|B_1)$.

3.6.4 Color Blindness

Some people believe that red-green color blindness is more prevalent in males than in females. To test this hypothesis we gather a random sample form the adult US population. Let M be the event a person is male, F is the event the person is female, C is the event the person has red-green color blindness and C' is the event he/she does not have red-green color blindness. From the data we obtain point estimates of the following probabilities: (1) $p_{11} = P(M \cap C) = 0.040$; (2) $p_{12} = P(M \cap C') = 0.493$; (3) $p_{21} = P(F \cap C) = 0.008$; and (4) $p_{22} = P(F \cap C') = 0.459$. The uncertainty in these point estimates will be shown in their fuzzy values : (1) $\bar{p}_{11} = (0.02/0.04/0.06)$; (2) $\bar{p}_{21} = (0.463/0.493/0.523)$; (3) $\bar{p}_{21} = (0.005/0.008/0.011)$; and (4) $\bar{p}_{22} = (0.439/0.459/0.479)$.

We wish to calculate the fuzzy conditional probabilities $\overline{P}(M|C)$ and

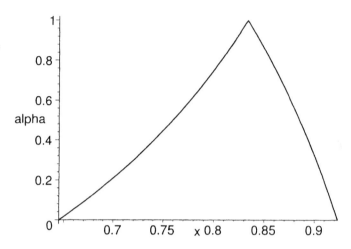

Figure 3.5: Fuzzy Probability of Male Given Color Blind

$\overline{P}(F|C)$. The α-cuts of the first fuzzy probability are

$$\overline{P}(M|C)[\alpha] = \{ \frac{p_{11}}{p_{21} + p_{11}} | \ \ \mathbf{S} \ \ \}, \qquad (3.41)$$

for $\alpha \in [0,1]$ and $\mathbf{S}$ denotes "$p_{ij} \in p_{ij}[\alpha]$, $1 \le i, j \le 2$ and $p_{11} + ... + p_{22} = 1$".
Let $H(p_{11}, p_{21}) = \frac{p_{11}}{p_{11} + p_{21}}$. Then H is an increasing function of p_{11} but
decreasing in p_{21}. So, as in the previous application, we obtain

$$\overline{P}(M|C)[\alpha] = [H(p_{111}(\alpha), p_{212}(\alpha)), H(p_{112}(\alpha), p_{211}(\alpha))], \qquad (3.42)$$

for all α. This fuzzy conditional probability is shown in Figure 3.5.
    Alpha-cuts of the second fuzzy conditional probability are

$$\overline{P}(F|C)[\alpha] = \{ \frac{p_{21}}{p_{21} + p_{11}} | \ \ \mathbf{S} \ \ \}, \qquad (3.43)$$

for all α. Let $G(p_{11}, p_{21}) = \frac{p_{21}}{p_{21} + p_{11}}$. Then G increases in p_{21} but decreases
in p_{11}. We may check that the following result is "feasible".

$$\overline{P}(F|C)[\alpha] = [G(p_{112}(\alpha), p_{211}(\alpha)), G(p_{111}(\alpha), p_{212}(\alpha))], \qquad (3.44)$$

for all α. This fuzzy probability is in Figure 3.6.
    Do we obtain $\overline{P}(F|C) \le \overline{P}(M|C)$. where here $\le$ means "less that or
equal to"?

3.6.5 Fuzzy Bayes

This is a numerical application showing how a fuzzy probability $\overline{P}(A_1)$
changes from the fuzzy prior to the fuzzy posterior. Let there be only two

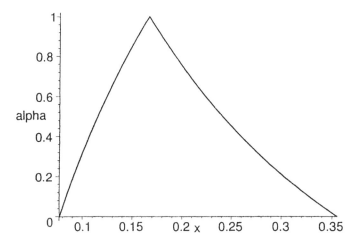

Figure 3.6: Fuzzy Probability of Female Given Color Blind

states of nature S_1 and S_2 with fuzzy prior probabilities $\bar{p}_1 = \overline{P}(S_1) = (0.3/0.4/0.5)$ and $\bar{p}_2 = \overline{P}(S_2) = (0.5/0.6/0.7)$. There are also only two events A_1 and A_2 in the partition of X with known conditional probabilities $p_{11} = P(A_1|S_1) = 0.2$, $p_{21} = P(A_2|S_1) = 0.8$, $p_{12} = P(A_1|S_2) = 0.7$ and $p_{22} = P(A_2|S_2) = 0.3$. We first find the fuzzy probability $\overline{P}(A_1)$ using the fuzzy prior probabilities. Its α-cuts are

$$\overline{P}(A_1)[\alpha] = \{(0.2)p_1 + (0.7)p_2| \quad \mathbf{S} \quad \}, \tag{3.45}$$

where $\mathbf{S}$ is " $p_i \in \bar{p}_i[\alpha]$, $1 \leq i \leq 2$, and $p_1 + p_2 = 1$". We easily evaluate this equation (3.45) and get the triangular fuzzy number $\overline{P}(A_1) = (0.41/0.50/0.59)$.

Now suppose we have information that event A_1 will occur. We need to obtain $\overline{P}(S_1|A_1)$ and $\overline{P}(S_2|A_1)$ from fuzzy Bayes' formula in Section 3.5. We first compute

$$\overline{P}(S_1|A_1)[\alpha] = \{\frac{(0.2)p_1}{(0.2)p_1 + (0.7)p_2}| \quad \mathbf{S} \quad \}, \tag{3.46}$$

and

$$\overline{P}(S_2|A_1)[\alpha] = \{\frac{(0.7)p_2}{(0.2)p_1 + (0.7)p_2}| \quad \mathbf{S} \quad \} \tag{3.47}$$

where $\mathbf{S}$ is "$p_i \in \bar{p}_i[\alpha]$, $i = 1, 2$ and $p_1 + p_2 = 1$". Both α-cuts are easily found. Let $\bar{p}_i[\alpha] = [p_{i1}(\alpha), p_{i2}(\alpha)]$, for $i = 1, 2$. Then

$$\overline{P}(S_1|A_1)[\alpha] = [\frac{0.2p_{11}(\alpha)}{0.2p_{11}(\alpha) + 0.7p_{22}(\alpha)}, \frac{0.2p_{12}(\alpha)}{0.2p_{12}(\alpha) + 0.7p_{21}(\alpha)}], \tag{3.48}$$

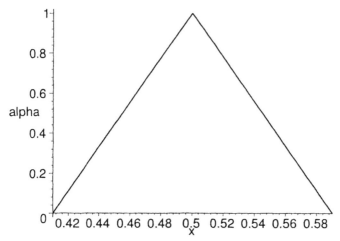

Figure 3.7: $\overline{P}(A_1)$ Using the Fuzzy Prior

and

$$\overline{P}(S_2|A_1)[\alpha] = [\frac{0.7p_{21}(\alpha)}{0.2p_{12}(\alpha) + 0.7p_{21}(\alpha)}, \frac{0.7p_{22}(\alpha)}{0.2p_{11}(\alpha) + p_{22}(\alpha)}], \qquad (3.49)$$

for all α. We get

$$\overline{P}(S_1|A_1)[\alpha] = [\frac{0.06 + 0.02\alpha}{0.55 - 0.05\alpha}, \frac{0.10 - 0.02\alpha}{0.45 + 0.05\alpha}], \qquad (3.50)$$

and

$$\overline{P}(S_2|A_1)[\alpha] = [\frac{0.35 + 0.07\alpha}{0.45 + 0.05\alpha}, \frac{0.49 - 0.07\alpha}{0.55 - 0.05\alpha}]. \qquad (3.51)$$

Now we may compute $\overline{P}(A_1)$ using the fuzzy posterior probabilities. It has α-cuts

$$(0.2)\overline{P}(S_1|A_1)[\alpha] + (0.7)\overline{P}(S_2|A_1)[\alpha]. \qquad (3.52)$$

The graphs of $\overline{P}(A_1)$ using the fuzzy prior is in Figure 3.7 and $\overline{P}(A_1)$ from the fuzzy posterior is shown in Figure 3.8. The fuzzy number in Figure 3.7 is the triangular fuzzy number $(0.41/0.50/0.59)$. The fuzzy number in Figure 3.8 may look like a triangular fuzzy number, but it is not. The sides of the fuzzy number in Figure 3.8 are slightly curved and are not straight lines. Why does $\overline{P}(A_1)$ in Figure 3.8 lie to the right of $\overline{P}(A_1)$ shown in Figure 3.7?

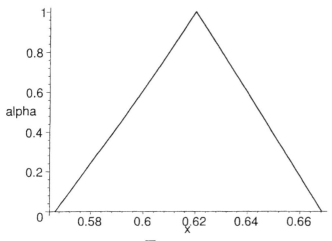

Figure 3.8: $\overline{P}(A_1)$ Using the Fuzzy Posterior

3.7 References

1. J.J.Buckley and E.Eslami: Uncertain Probabilities I: The Discrete Case, Soft Computing. To appear.

2. Maple 6, Waterloo Maple Inc. Waterloo, Canada.

Chapter 4

Discrete Fuzzy Random Variables

4.1 Introduction

This chapter is based on [1]. New material, not included in [1], includes the fuzzy Poisson (Section 4.3) and the applications in Section 4.4.

We start with the fuzzy binomial. The crisp binomial probability mass function, usually written $b(m, p)$ where m is the number of independent experiments and p is the probability of a "success" in each experiment, has one parameter p. We assume that p is not known exactly and is to be estimated from a random sample or from expert opinion. In either case the result is, justified in Section 2.8, that we substitute a fuzzy number $\overline{p}$ for p to get the fuzzy binomial. Then we discuss the fuzzy Poisson probability mass function. The crisp Poisson probability mass function has one parameter, usually written λ, which we also assume is not known exactly. Hence we substitute fuzzy number $\overline{\lambda}$ for λ to obtain the fuzzy Poisson probability mass function. The fuzzy binomial and the fuzzy Poisson comprises the next two sections. We look at some applications of these two discrete fuzzy probability mass functions in Section 4.4.

4.2 Fuzzy Binomial

As before $X = \{x_1, ..., x_n\}$ and let E be a non-empty, proper, subset of X. We have an experiment where the result is considered a "success" if the outcome x_i is in E. Otherwise, the result is considered a "failure". Let $P(E) = p$ so that $P(E') = q = 1 - p$. $P(E)$ is the probability of success and $P(E')$ is the probability of failure. We assume that $0 < p < 1$.

Suppose we have m independent repetitions of this experiment. If $P(r)$

is the probability of r successes in the m experiments, then

$$P(r) = \binom{m}{r} p^r q^{m-r}, \tag{4.1}$$

for $r = 0, 1, 2, ..., m$, gives the binomial distribution.

In these experiments let us assume that $P(E)$ is not known precisely and it needs to be estimated, or obtained from expert opinion. So the p value is uncertain and we substitute $\bar{p}$ for p and $\bar{q}$ for q so that there is a $p \in \bar{p}[1]$ and a $q \in \bar{q}[1]$ with $p + q = 1$. Now let $\bar{P}(r)$ be the fuzzy probability of r successes in m independent trials of the experiment. Under our restricted fuzzy algebra we obtain

$$\bar{P}(r)[\alpha] = \{ \binom{m}{r} p^r q^{m-r} | \quad \mathbf{S} \quad \}, \tag{4.2}$$

for $0 \leq \alpha \leq 1$, where now $\mathbf{S}$ is the statement "$p \in \bar{p}[\alpha], q \in \bar{q}[\alpha], p + q = 1$". Notice that $\bar{P}(r)$ is not $\binom{m}{r} \bar{p}^r \bar{q}^{m-r}$. If $\bar{P}(r)[\alpha] = [P_{r1}(\alpha), P_{r2}(\alpha)]$, then

$$P_{r1}(\alpha) = min\{ \binom{m}{r} p^r q^{m-r} | \quad \mathbf{S} \quad \}, \tag{4.3}$$

and

$$P_{r2}(\alpha) = max\{ \binom{m}{r} p^r q^{m-r} | \quad \mathbf{S} \quad \}. \tag{4.4}$$

Example 4.2.1

Let $p = 0.4$, $q = 0.6$ and $m = 3$. Since p and q are uncertain we use $\bar{p} = (0.3/0.4/0.5)$ for p and $\bar{q} = (0.5/0.6/0.7)$ for q. Now we will calculate the fuzzy number $\bar{P}(2)$. If $p \in \bar{p}[\alpha]$ then $q = 1 - p \in \bar{q}[\alpha]$. Equations (4.3) and (4.4) become

$$P_{r1}(\alpha) = min\{3p^2 q | \quad \mathbf{S} \quad \}, \tag{4.5}$$

and

$$P_{r2}(\alpha) = max\{3p^2 q | \quad \mathbf{S} \quad \}. \tag{4.6}$$

Since $d(3p^2(1 - p))/dp > 0$ on $\bar{p}[0]$ we obtain

$$\bar{P}(2)[\alpha] = [3(p_1(\alpha))^2 (1 - p_1(\alpha)), 3(p_2(\alpha))^2 (1 - p_2(\alpha))], \tag{4.7}$$

where $\bar{p}[\alpha] = [p_1(\alpha), p_2(\alpha)] = [0.3 + 0.1\alpha, 0.5 - 0.1\alpha]$.

Alpha-cuts of the fuzzy mean and the fuzzy variance of the fuzzy binomial distribution are calculated as in equations (3.12) and (3.13) in Chapter 3, respectively. In the crisp case we know $\mu = mp$ and $\sigma^2 = mpq$. Does $\bar{\mu} = m\bar{p}$

and $\bar{\sigma}^2 = m\bar{p} \cdot \bar{q}$? We now argue that the correct result is $\bar{\mu} \leq m\bar{p}$ and $\bar{\sigma}^2 \leq m\bar{p} \cdot \bar{q}$. We see that

$$\bar{\mu}[\alpha] = \{\sum_{r=1}^{m} r\binom{m}{r} p^r q^{m-r} | \quad \mathbf{S} \quad \},\qquad(4.8)$$

which simplifies to

$$\bar{\mu}[\alpha] = \{mp| \quad \mathbf{S} \quad \}.\qquad(4.9)$$

Let $s \in \bar{\mu}[\alpha]$. Then $s = mp$ for $p \in \bar{p}[\alpha]$, $q \in \bar{q}[\alpha]$ and $p + q = 1$. Hence, $s \in m\bar{p}[\alpha]$. So $\bar{\mu} \leq m\bar{p}$. To show they may not be equal let $\bar{p} = (0.2/0.3/0.4)$ and $\bar{q} = (0.65/0.7/0.75)$. Then $\bar{\mu}[0] = m[0.25, 0.35]$ but $m\bar{p}[0] = m[0.2, 0.4]$. If $\bar{q} = 1 - \bar{p}$, then $\bar{\mu} = m\bar{p}$.

To show $\bar{\sigma}^2 \leq m\bar{p} \cdot \bar{q}$ we see first, as we get equation (4.9) from equation (4.8), that

$$\bar{\sigma}^2[\alpha] = \{mpq| \quad \mathbf{S} \quad \}.\qquad(4.10)$$

Then we argue, just like before, that given $s \in \bar{\sigma}^2[\alpha]$, then $s \in m\bar{p}[\alpha]\bar{q}[\alpha]$. This shows $\bar{\sigma}^2 \leq m\bar{p} \cdot \bar{q}$. Now, to show that they may not be equal let $\bar{p}$ and $\bar{q}$ be given as above. Then $m\bar{p}[0]\bar{q}[0] = m[0.13, 0.30]$ but $\bar{\sigma}^2[0] = m[(0.25)(0.75), (0.35)(0.65)] = m[0.1875, 0.2275]$.

Example 4.2.2

We may find the α-cuts of $\bar{\sigma}^2$ if $\bar{q} = 1 - \bar{p}$. Let $\bar{p} = (0.4/0.6/0.8)$ and $\bar{q} = (0.2/0.4/0.6)$. Then

$$\bar{\sigma}^2[\alpha] = \{mp(1-p)|p \in \bar{p}[\alpha]\},\qquad(4.11)$$

from equation (4.10). Let $h(p) = mp(1-p)$. We see that $h(p) :$ (1) is increasing on $[0, 0.5]$; (2) has its maximum of $0.25m$ at $p = 0.5$; and (3) is decreasing on $[0.5, 1]$. So, the evaluation of equation (4.11), see Section 2.9, depends if $p = 0.5$ belongs to the α-cut of $\bar{p}$. Let $\bar{p}[\alpha] = [p_1(\alpha), p_2(\alpha)] = [0.4 + 0.2\alpha, 0.8 - 0.2\alpha]$. So, $p = 0.5$ belongs to the α-cut of $\bar{p}$ only for $0 \leq \alpha \leq 0.5$. Then

$$\bar{\sigma}^2[\alpha] = [h(p_2(\alpha)), 0.25m],\qquad(4.12)$$

for $0 \leq \alpha \leq 0.5$, and

$$\bar{\sigma}^2[\alpha] = [h(p_2(\alpha)), h(p_1(\alpha))],\qquad(4.13)$$

for $0.5 \leq \alpha \leq 1$. We substitute in for $p_1(\alpha)$ and $p_2(\alpha)$ to finally obtain $\bar{\sigma}^2$ and its graph, for $m = 10$, is in Figure 4.1.

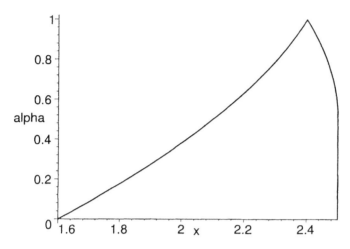

Figure 4.1: Fuzzy Variance in Example 4.2.2

4.3   Fuzzy Poisson

Let X be a random variable having the Poisson probability mass function. If $P(x)$ stands for the probability that $X = x$, then

$$P(x) = \frac{\lambda^x \exp(-\lambda)}{x!}, \qquad (4.14)$$

for $x = 0, 1, 2, 3, ...$, and parameter $\lambda > 0$. Now substitute fuzzy number $\overline{\lambda} > 0$ for λ to produce the fuzzy Poisson probability mass function. Set $\overline{P}(x)$ to be the fuzzy probability that $X = x$. Then we find α-cuts of this fuzzy number as

$$\overline{P}(x)[\alpha] = \{ \frac{\lambda^x \exp(-\lambda)}{x!} | \lambda \in \overline{\lambda}[\alpha] \}, \qquad (4.15)$$

for all $\alpha \in [0, 1]$. The evaluation of equation (4.15) depends on the relation of x to $\overline{\lambda}[0]$. Let $h(\lambda) = \frac{\lambda^x \exp(-\lambda)}{x!}$ for fixed x and $\lambda > 0$. We see that $h(\lambda)$ is an increasing function of λ for $\lambda < x$, the maximum value of $h(\lambda)$ occurs at $\lambda = x$, and $h(\lambda)$ is a decreasing function of λ for $\lambda > x$. Let $\overline{\lambda}[\alpha] = [\lambda_1(\alpha), \lambda_2(\alpha)]$, for $0 \leq \alpha \leq 1$. Then we see that : (1) if $\lambda_2(0) < x$, then $\overline{P}(x)[\alpha] = [h(\lambda_1(\alpha)), h(\lambda_2(\alpha))]$; and (2) if $x < \lambda_1(0)$, then $\overline{P}(x)[\alpha] = [h(\lambda_2(\alpha)), h(\lambda_1(\alpha))]$. The other case, where $x \in \overline{\lambda}[0]$, is explored in the following example.

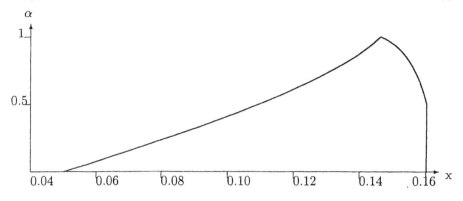

Figure 4.2: Fuzzy Probability in Example 4.3.1

Example 4.3.1

Let $x = 6$ and $\overline{\lambda} = (3/5/7)$. We see that $x \in [3,7] = \overline{\lambda}[0]$. We determine $\overline{\lambda}[\alpha] = [3 + 2\alpha, 7 - 2\alpha]$. Define $\overline{P}(6)[\alpha] = [p_1(\alpha), p_2(\alpha)]$. To determine the α-cuts of $\overline{P}(6)$ we need to solve (see equations (2.24) and (2.25) in Chapter 2)

$$p_1(\alpha) = min\{h(\lambda)|\lambda \in \overline{\lambda}[\alpha]\}, \qquad (4.16)$$

and

$$p_2(\alpha) = max\{h(\lambda)|\lambda \in \overline{\lambda}[\alpha]\}. \qquad (4.17)$$

It is not difficult to solve equations (4.16) and (4.17) producing

$$\overline{P}(6)[\alpha] = [h(3 + 2\alpha), h(6)], \qquad (4.18)$$

for $0 \le \alpha \le 0.5$, and

$$\overline{P}(6)[\alpha] = [h(3 + 2\alpha), h(7 - 2\alpha)], \qquad (4.19)$$

for $0.5 \le \alpha \le 1$. The graph of $\overline{P}(6)$ is shown in Figure 4.2.

Let us consider another, slightly more complicated, example of finding fuzzy probabilities using the fuzzy Poisson.

Example 4.3.2

Let $\overline{\lambda} = (8/9/10)$ and define $\overline{P}([3, \infty))$ to be the fuzzy probability that $X \ge 3$. Also let $\overline{P}([3, \infty))[\alpha] = [q_1(\alpha), q_2(\alpha)]$. Then

$$q_1(\alpha) = min\{1 - \sum_{x=0}^{2} h(\lambda)|\lambda \in \overline{\lambda}[\alpha]\}, \qquad (4.20)$$

and

$$q_2(\alpha) = max\{1 - \sum_{x=0}^{2} h(\lambda)|\lambda \in \overline{\lambda}[\alpha]\}, \qquad (4.21)$$

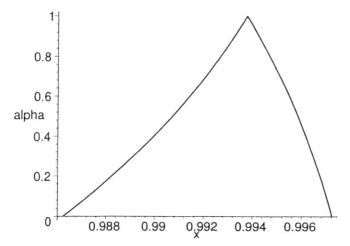

Figure 4.3: Fuzzy Probability in Example 4.3.2

for all α. Let $k(\lambda) = 1 - [\sum_{x=0}^{2} h(\lambda)]$. Then $dk/d\lambda > 0$ for $\lambda > 0$ Hence, we may evaluate equations (4.20) and (4.21) and get

$$\overline{P}([3, \infty))[\alpha] = [k(\lambda_1(\alpha)), k(\lambda_2(\alpha))]. \tag{4.22}$$

This fuzzy probability is shown in Figure 4.3.

To finish this section we now compute the fuzzy mean and the fuzzy variance of the fuzzy Poisson probability mass function. Alpha-cuts of the fuzzy mean, from equation (3.12) of Chapter 3, are

$$\overline{\mu}[\alpha] = \{\sum_{x=0}^{\infty} xh(\lambda) | \lambda \in \overline{\lambda}[\alpha]\}, \tag{4.23}$$

which reduces to, since the mean of the crisp Poisson is λ, the expression

$$\overline{\mu}[\alpha] = \{\lambda | \lambda \in \overline{\lambda}[\alpha]\}. \tag{4.24}$$

Hence, $\overline{\mu} = \overline{\lambda}$. So the fuzzy mean is just the fuzzification of the crisp mean. Let the fuzzy variance be $\overline{\sigma}^2$ and we obtain its α-cuts as

$$\overline{\sigma}^2[\alpha] = \{\sum_{x=0}^{\infty} (x - \mu)^2 h(\lambda) | \lambda \in \overline{\lambda}[\alpha], \mu = \lambda\}, \tag{4.25}$$

which reduces to, since the variance of the crisp Poisson is also λ, the expression

$$\overline{\sigma}^2[\alpha] = \{\lambda | \lambda \in \overline{\lambda}[\alpha]\}. \tag{4.26}$$

It follows that $\overline{\sigma}^2 = \overline{\lambda}$ and the fuzzy variance is the fuzzification of the crisp variance.

4.4 Applications

In this section we look at three applications: (1) using the fuzzy Poisson to approximate values of the fuzzy binomial; (2) using the fuzzy binomial to calculate the fuzzy probabilities of "overbooking"; and (3) using then fuzzy Poisson to estimate the size of a rapid response team to terrorist attacks.

4.4.1 Fuzzy Poisson Approximating Fuzzy Binomial

Let X be a random variable having the binomial probability mass function $b(n,p)$. From crisp probability theory [2] we know that if n is large and p is small we can use the Poisson to approximate values of the binomial. For non-negative integers a and b, $0 \leq a \leq b$, let $P([a,b])$ be the probability that $a \leq X \leq b$. Then using the binomial we have ($q = 1 - p$)

$$P([a,b]) = \sum_{x=a}^{b} \binom{n}{x} p^x q^{n-x}. \tag{4.27}$$

Using the Poisson, with $\lambda = np$, we calculate

$$P([a,b]) \approx \sum_{x=a}^{b} \frac{\lambda^x \exp(-\lambda)}{x!}. \tag{4.28}$$

Now switch to the fuzzy case. Let $\overline{p}$ be small, which means that all $p \in \overline{p}[0]$ are sufficiently small. Let $\overline{P}([a,b])$ be the fuzzy probability that $a \leq X \leq b$. For notational simplicity set $\overline{P}_{b\alpha} = \overline{P}([a,b])[\alpha]$ using the fuzzy binomial. Also set $\overline{P}_{p\alpha} = \overline{P}([a,b])[\alpha]$ using the fuzzy Poisson approximation. Then

$$\overline{P}_{b\alpha} = \{\sum_{x=a}^{b} \binom{n}{x} p^x (1-p)^{n-x} | p \in \overline{p}[\alpha]\}, \tag{4.29}$$

and

$$\overline{P}_{p\alpha} = \{\sum_{x=a}^{b} \frac{\lambda^x \exp(-\lambda)}{x!} | \lambda \in n\overline{p}[\alpha]\}. \tag{4.30}$$

Notice that in equation (4.29) we are using a slightly different model of the fuzzy binomial from equation (4.2) which is similar to, but not exactly equal to, $\overline{q} = 1 - \overline{p}$. We now argue that $\overline{P}_{b\alpha} \approx \overline{P}_{p\alpha}$ for all α. This approximation

α	$\overline{P}_{b\alpha}$	$\overline{P}_{p\alpha}$
0	[0.647,0.982]	[0.647,0.981]
0.2	[0.693,0.967]	[0.692,0.966]
0.4	[0.737,0.948]	[0.736,0.946]
0.6	[0.780,0.923]	[0.779,0.921]
0.8	[0.821,0.893]	[0.819,0.891]
1.0	0.859	0.857

Table 4.1: Fuzzy Poisson Approximation to Fuzzy Binomial

is to be interpreted as follows: (1) given $z \in \overline{P}_{b\alpha}$, there is a $y \in \overline{P}_{p\alpha}$ so that $z \approx y$; and (2) given $y \in \overline{P}_{p\alpha}$ there is a $z \in \overline{P}_{b\alpha}$ so that $y \approx z$. Also, $z \approx y$ and $y \approx z$ are to be interpreted as in crisp probability theory. To show (1) let $z \in \overline{P}_{b\alpha}$, then $z = \sum_{x=a}^{b} \binom{n}{x} p^x (1-p)^{n-x}$ for some $p \in \overline{p}[\alpha]$. For this same p let $\lambda = np$ and set $y = \sum_{x=a}^{b} \frac{\lambda^x \exp(-\lambda)}{x!}$. Then $z \approx y$. Similarly we show (2).

Example 4.4.1.1

Let $n = 100$ and $p = 0.02$. Then set $\overline{p} = (0.01/0.02/0.03)$. Now let $a = 0$ and $b = 3$ so $\overline{P}_{b\alpha} = \overline{P}([0,3])[\alpha]$ using the fuzzy binomial and $\overline{P}_{p\alpha} = \overline{P}([0,3])[\alpha]$ using the fuzzy Poisson approximation. We have computed values of $\overline{P}_{b\alpha}$ and $\overline{P}_{p\alpha}$ for $\alpha = 0, 0.2, 0.4, 0.6, 0.8, 1$ and these are shown in Table 4.1.

To compute $\overline{P}_{b\alpha}$ we simply graphed the function $F(p) = \sum_{x=0}^{3} \binom{n}{x} p^x (1-p)^{n-x}$ for p in the interval $\overline{p}[\alpha]$, using the software package Maple [3], to pick out the end points of the α-cut. It turns out that $F(p)$ is a decreasing function of p over the interval $\overline{p}[0]$.

Computing $\overline{P}_{p\alpha}$ was easier. Let $G(\lambda) = \sum_{x=0}^{3} \frac{\lambda^x \exp(-\lambda)}{x!}$. We see that $dG/d\lambda < 0$ so if $\overline{\lambda}[\alpha] = [\lambda_1(\alpha), \lambda_2(\alpha)]$, then $\overline{P}_{p\alpha} = [G(\lambda_2(\alpha)), G(\lambda_1(\alpha))]$. Here we use $\overline{\lambda} = n\overline{p} = (1/2/3)$.

4.4.2   Overbooking

Americana Air has the policy of booking as many as 120 persons on an airplane that can seat only 114. Past data implies that approximately only 85% of the booked passengers actually arrive for the flight. We want to find the probability that if Americana Air books 120 persons, not enough seats will be available.

This is a binomial situation with $p \approx 0.85$. Since p has been estimated from past data we use a set of confidence intervals, see Section 2.8, to construct a fuzzy number $\overline{p} = (0.75/0.85/0.95)$ for p producing the fuzzy binomial. Let $\overline{P}_0$ be the fuzzy probability of being overbooked, then its α-cuts

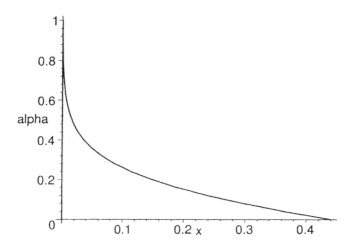

Figure 4.4: Fuzzy Probability of Overbooking

are

$$\overline{P}_0[\alpha] = \{ \sum_{x=115}^{120} \binom{120}{x} p^x (1-p)^{120-x} | p \in \overline{p}[\alpha] \}. \qquad (4.31)$$

Again, as in the previous subsection, we are using a slightly different form of the fuzzy binomial than given in Section 4.2. The graph of the fuzzy probability of overbooking is shown in Figure 4.4. Let $F(p) = \sum_{x=115}^{120} \binom{120}{x} p^x (1 - p)^{120-x}$ for $p \in \overline{p}[0]$. We graphed $F(p)$ using Maple and found that this function is an increasing function of p on the interval $\overline{p}[0]$. This made it easy to evaluate equation (4.31) and obtain the graph in Figure 4.4. Figure 4.4 does not show the left side of the fuzzy number because the left side of the α-cuts involve very small numbers. Selected α-cuts of $\overline{P}_0$ are: (1) $[0.9(10)^{-9}, 0.4415]$ for $\alpha = 0$; (2) $[0.5(10)^{-6}, 0.0160]$ for $\alpha = 0.5$; and (3) $[0.00014, 0.00014]$ for $\alpha = 1$.

Notice that the core of $\overline{P}_0$, where the membership is one, is just the crisp probability of overbooking using $p = 0.85$. The spread of the fuzzy number $\overline{P}_0$ shows the uncertainty about the crisp result.

4.4.3   Rapid Response Team

The US government is planning a rapid response team to terrorist attacks within continental US. They need to compute the probability of multiple

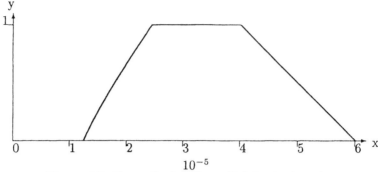

Figure 4.5: Fuzzy Probability of Multiple Attacks

attacks in a single day to see if they will need one team or multiple teams. It is difficult to do, but they estimate that the mean number of terrorist attacks per day is approximately $\lambda = 0.008$, or about 3 per year starting in 2003. Using the Poisson probability mass function, find the probability that the number of attacks in one day is 0, or 1, or at least 2.

The value of λ was estimated by a group of experts and is very uncertain. Hence we will use a fuzzy number $\overline{\lambda} = (0.005/0.007, 0.009/0.011)$, a trapezoidal fuzzy number, for λ. Let $\overline{P}_m$ be the fuzzy probability of 2 or more attacks per day, which will be used to see if multiple rapid response teams will be needed. Alpha-cuts of this fuzzy probability are

$$\overline{P}_m[\alpha] = \{1 - \sum_{x=0}^{1} \frac{\lambda^x \exp(-\lambda)}{x!} | \lambda \in \overline{\lambda}[\alpha]\}, \qquad (4.32)$$

for $0 \le \alpha \le 1$. Let $v(\lambda) = 1 - \sum_{x=0}^{1} \frac{\lambda^x \exp(-\lambda)}{x!}$. We find that $dv/d\lambda > 0$ so if $\overline{\lambda}[\alpha] = [\lambda_1(\alpha), \lambda_2(\alpha)]$, then

$$\overline{P}_m[\alpha] = [v(\lambda_1(\alpha)), v(\lambda_2(\alpha))]. \qquad (4.33)$$

The graph of $\overline{P}_m$ is in Figure 4.5. Notice that this fuzzy number is a trapezoidal shaped fuzzy number. Do they need multiple rapid response teams?

4.5   References

1. J.J.Buckley and E.Eslami: Uncertain Probabilities I: The Discrete Case, Soft Computing. To appear.

2. R.V.Hogg and E.A.Tanis: Probability and Statistical Inference, Sixth Edition, Prentice Hall, Upper Saddle River, N.J., 2001.

3. Maple 6, Waterloo Maple Inc., Waterloo, Canada.

Chapter 5

Fuzzy Queuing Theory

5.1 Introduction

The results in Section 5.3 come from [1]. However, the applications in Sections 5.4 were adapted from [2]. We first review some basic results from regular, finite, Markov chains (also in the next Chapter) applied to queuing theory and then show in Section 5.3 that we also obtain steady state fuzzy probabilities for fuzzy regular, finite, Markov chains. Then in Section 5.4 we discuss two applications.

5.2 Regular, Finite, Markov Chains

We will consider only one queuing system in this Chapter. The reader may then extend these results to other queuing systems. Almost any book on operations research (management science) contains a chapter covering the basics of queuing theory. In this section, and the following section, the system we will look at is : c parallel, and identical, servers; finite (M) system capacity; and infinite calling source (where the customers come from). We will model the system as a regular, finite, Markov chain. So let us first briefly review the needed basic results from finite, and regular, Markov chains.

A finite Markov chain has a finite number of possible states (outcomes) $S_1, S_2, \ldots, S_r$ at each step $n = 1, 2, 3 \ldots$, in the process. Let

$$p_{ij} = Prob(S_j \ at \ step \ n+1 | S_i \ at \ step \ n), \qquad (5.1)$$

$1 \leq i, j \leq r$, $n = 1, 2, \ldots$. The p_{ij} are the transition probabilities which do not depend on n. The transition matrix $P = (p_{ij})$ is a $r \times r$ matrix of the transition probabilities. An important property of P is that the row sums are equal to one and each $p_{ij} \geq 0$. Let $p_{ij}^{(n)}$ be the probability of starting in state S_i and ending up in S_j after n steps. Define P^n to be the product of P n–times and it is well known that $P^n = (p_{ij}^{(n)})$ for all n. If $p^{(0)} = (p_1^{(0)}, \ldots, p_r^{(0)})$, where $p_i^{(0)} =$ the probability of initially being in state S_i, and $p^{(n)} = (p_1^{(n)}, \cdots, p_r^{(n)})$, where $p_i^{(n)} =$ the probability of being in state S_i after n steps, we know that $p^{(n)} = p^{(0)} P^n$.

We say that the Markov chain is regular if $P^k > 0$ for some k, which is $p_{ij}^{(k)} > 0$ for all i, j. This means that it is possible to go from any state S_i to any state S_j in k steps. A property of regular Markov chains is that powers of P converge, or $\lim_{n \to \infty} P^n = \Pi$, where the rows of Π are identical. Let w be the unique left eigenvalue of P corresponding to eigenvalue one, so that $w_i > 0$ all i and $\sum_{i=1}^r w_i = 1$. That is $wP = w$ for $1 \times r$ vector w. Each row in Π is equal to w and $p^{(n)} \to p^{(0)} \Pi = w$. After a long time, thinking that each step being a time interval, the probability of being in state S_i is w_i, $1 \le i \le r$, independent of the initial conditions $p^{(0)}$. In a regular Markov chain the process goes on forever jumping from state to state, to state, ...

Now we can consider our queuing system with c parallel and identical servers, system capacity $M > c$ and an infinite calling source. The calling source is the pool of potential customers. We assume the system changes can occur only at the end of a time interval δ. This time interval may be one second, one minute, one hour, etc. During a time interval δ: (1) customers may arrive at the system but are only allowed into the system at the end of the time interval; (2) customers may leave the servers but are allowed to return to the calling source only at the end of the time interval; (3) at the end of the time interval all customers in queue (in the system but not in the servers) are allowed to fill the empty servers; and (4) all customers who arrived are allowed into the system to fill empty servers or go into queue up to capacity M with all others turned away to return to the calling source. System changes can occur only at times $t = \delta, 2\delta, 3\delta, \ldots$

Let $p(i)$ be the probability that i customers arrive at the system during a time interval δ, $i = 0, 1, 2, 3\ldots$ Then $\sum_{i=0}^{\infty} p(i) = 1$. Next let $q(l, s)$ be the probability that , during a time interval δ, l customers in the servers complete service and are waiting to return to the calling source at the end of the time interval , given that s servers are full of customers at the start of the time period, for $l = 0, 1, 2, \ldots, s$ and $s = 0, 1, 2, 3, \ldots c$. Then $\sum_{l=0}^{s} q(l, s) = 1$ for each s. Next we construct the transition matrix P. The rows of P are labeled $0, 1, 2, 3, \ldots, M$ representing the state of the system at the start of the time period and the columns of P are labeled $0, 1, 2, 3, \ldots, M$ representing the state of the system at the beginning of the next period. To see how we compute the p_{ij} in P let us look at the following example.

Example 5.2.1

Let $c = 2$ and $M = 4$. Then $P = (p_{ij})$, a 5×5 matrix with probabilities p_{ij}, $0 \le i, j \le 4$. Let us compute the p_{2j} in the third row: (1)

$p_{20} = p(0)q(2,2)$; (2) $p_{21} = p(0)q(1,2) + p(1)q(2,2)$; (3) $p_{22} = p(0)q(0,2) + p(1)q(1,2) + p(2)q(2,2)$; (4) $p_{23} = p(1)q(0,2) + p(2)q(1,2) + p(3)q(2,2)$; and (5) $p_{24} = p^*(2)q(0,2) + p^*(3)q(1,2) + p^*(4)q(2,2)$ where $p^*(k) = \sum_{i=k}^{\infty} p(i)$. Notice that $\sum_{j=0}^{4} p_{2j} = 1$ so P is the transition matrix for a finite Markov chain. Also, $P^2 > 0$ so it is regular.

We see from Example 5.2.1 that P will be the transition matrix for a regular, finite, Markov chain. Then $P^n \rightarrow \Pi$ where each row in Π is $w = (w_0, ..., w_M)$ and $w_i > 0$, $w_0 + ... + w_M = 1$, $wP = w$. Also $p^{(n)} \rightarrow w$ so, after a long time, the probability of being in state S_j is independent of the initial conditions.

5.3 Fuzzy Queuing Theory

The $p(i)$ and the $q(l,s)$ have to be known exactly. Many times they have to be estimated and if the queuing system is in the planning stage, these probabilities will be estimated by experts. So we assume there is uncertainty in some of the $p(i)$ and $q(l,s)$ values which implies uncertainty in the p_{ij} values in P. If a $p_{ij} = 0$ in P, then we assume there is no uncertainty in this value. We model the uncertainty by substituting fuzzy numbers $\overline{p}_{ij}$ for all the non-zero p_{ij} in P. Of course. if some p_{ij} value is known, say $p_{24} = 0.15$ with no uncertainty, then $\overline{p}_{24} = 0.15$ also. We then obtain a fuzzy transition matrix $\overline{P} = (\overline{p}_{ij})$ so that if $p_{ij} = 0$ then $\overline{p}_{ij} = 0$ and if $0 < p_{ij} < 1$, then $0 < \overline{p}_{ij} < 1$ also. We now argue, under restricted fuzzy matrix multiplication, that $\overline{P}^n \rightarrow \overline{\Pi}$ where each row in $\overline{\Pi}$ is $\overline{w} = (\overline{w}_0, ..., \overline{w}_M)$ and the $\overline{w}_i$ give the fuzzy steady state probabilities of the fuzzy queuing system.

The uncertainty is in some of the p_{ij} values but not in the fact that the rows in the transition matrix must be discrete probability distributions (row sums equal one). So we now put the following restriction on the $\overline{p}_{ij}$ values: there are $p_{ij} \in \overline{p}_{ij}[1]$ so that $P = (p_{ij})$ is the transition matrix for a finite Markov chain (row sums one). We will need the following definitions for our restricted fuzzy matrix multiplication. Define

$$Dom_i[\alpha] = (\prod_{j=0}^{M} \overline{p}_{ij}[\alpha]) \bigcap S, \qquad (5.2)$$

where " S", in equation (5.2), was defined in equation (3.2) in Chapter 3, for $0 \leq \alpha \leq 1$ and $0 \leq i \leq M$. Then set

$$Dom[\alpha] = \prod_{i=0}^{M} Dom_i[\alpha]. \qquad (5.3)$$

Define $v = (p_{00}, p_{01}, ..., p_{MM})$. For each $v \in Dom[\alpha]$ set $P = (p_{ij})$ and we get $P^n \rightarrow \Pi$. Let $\Gamma(\alpha) = \{w | wP = w, v \in Dom[\alpha]\}$. $\Gamma(\alpha)$ consists of all vectors w, which are the rows in Π, for all $v \in Dom[\alpha]$. Now the rows in $\overline{\Pi}$ will be all the same so let $\overline{w} = (\overline{w}_0...., \overline{w}_M)$ be a row in $\overline{\Pi}$. Also, let $\overline{w}_j[\alpha] = [w_{j1}(\alpha), w_{j2}(\alpha)]$, for $0 \leq j \leq M$. Then

$$w_{j1}(\alpha) = min\{w_j | w \in \Gamma(\alpha)\}, \qquad (5.4)$$

and

$$w_{j2}(\alpha) = max\{w_j | w \in \Gamma(\alpha)\}, \qquad (5.5)$$

where w_j is the jth component in the vector w. The steady state fuzzy probabilities are : (1) $\overline{w}_0 =$ the fuzzy probability of the system being empty; (2) $\overline{w}_1 =$ the fuzzy probability of one customer in the system; etc.

   In general, the solutions to equations (5.4) and (5.5) will be computationally difficult and one might consider using a genetic, or evolutionary, algorithm to get approximate solutions. However, in certain simple cases we can get the $\overline{w}_i$ by hand calculation.

Example 5.3.1

Assume that $c = 1$ and $M = 2$. We know the $p(i)$ but not the $q(i, s)$ since we are installing a new server. Let $p(0) = 0.3$, $p(1) = 0.2$, $p(2) = 0.1$, $q(0, 1) \approx 0.7$ and $q(1, 1) \approx 0.3$. To model this uncertainty let $\overline{q}(0, 1) = (0.6/0.7/0.8)$ and $\overline{q}(1, 1) = (0.2/0.3/0.4)$. The crisp transition matrix P is 3×3 with : (1) $p_{00} = 0.3$, $p_{01} = 0.2$ and $p_{02}(2) = p^*(2) = 0.5$; (2) $p_{10} = 0.3q(1, 1)$, $p_{11} = 0.3q(0, 1) + 0.2q(1, 1)$, $p_{12} = p^*(2)q(1, 1) + p^*(1)q(0, 1) = 0.5q(1, 1) + 0.7q(0, 1)$; and (3) $p_{20} = 0$, $p_{21} = 0.3q(1, 1)$, $p_{22} = 1.0q(0, 1) + p^*(1)q(1, 1) = q(0, 1) + 0.7q(1, 1)$. We may solve $wP = w$, $w_i > 0$, $w_0 + w_1 + w_2 = 1$ for the w_i in terms of the $q(0, 1)$ and $q(1, 1)$. For notational simplicity let $Y = q(0, 1)$ and $Z = q(1, 1)$, but $Y + Z = 1$ so we will use $Y = 1 - Z$ for Y. Then $w_0 = (0.9)Z^2/T$, $w_1 = 2.1Z/T$, and $w_2 = [4.9 + 0.7Z - 0.18Z^2]/T$ for $T = [4.9 + 2.8Z + 0.72Z^2]$. We find that: (1) w_0 is an increasing function of Z; (2) w_1 is also an increasing function of Z; but (3) w_2 is a decreasing function of Z. This enables us to solve equations (5.4) and (5.5) as Z varies in the α-cut of $\overline{q}(1, 1)$. For example, to find the end points of the interval $\overline{w}_2[\alpha]$ we: (1) use $Z = 0.2 + 0.1\alpha$, the left end point of $\overline{q}(1, 1) = (0.2/0.3/0.4)$, in the expression for w_2 a function of Z, to obtain the right end point of the interval $\overline{w}_2[\alpha]$; and (2) use $Z = 0.4 - 0.1\alpha$ in the function for w_2 to get the left end point of $\overline{w}_2[\alpha]$. The result gives triangular shaped fuzzy numbers for the $\overline{w}_i$ whose $\alpha = 0$ and $\alpha = 1$ cuts are shown in Table 5.1. A triangular shaped fuzzy number has curves for the sides of the triangle instead of straight lines.

5.4   Applications

In this section we present two applications of fuzzy queuing theory. Both applications are taken from [2], however the models are quite different. In

	$\alpha = 0$	$\alpha = 1$
$\overline{w}_0$	[0.0066,0.0235]	0.0140
$\overline{w}_1$	[0.0765,0.1369]	0.1085
$\overline{w}_2$	[0.8396,0.9169]	0.8775

Table 5.1: Alpha-cuts of the Fuzzy Probabilities in Example 5.3.1

[2] we used possibility theory and now we are using fuzzy probabilities. In the first application the system is: c parallel, and identical, servers; finite system capacity N; and finite calling source N. In the second application the system is: c parallel, and identical, servers; finite system capacity M; and finite calling source $N > M \geq c$.

5.4.1 Machine Servicing Problem

There are c repair crews available to service N machines. Let R be the hourly cost of each repair crew and let B be the hourly cost of lost production per waiting machine. When a machine breaks down it is assigned a repair crew, if one is available, or it has to wait for the first free repair crew. To model this problem with a fuzzy transition matrix $\overline{P}$ we assume repair crews are scheduled hourly and repaired machines go back into production on the hour. If this seems unreasonable, then we could take 30 minutes, or 5 minutes, for our unit of time.

The cost function, which is to be minimized, as a function of $c = 1, 2, ..., N$, is

$$\overline{Z}(c) = Rc + \overline{B}\{Expected[\overline{w}_c]\}, \qquad (5.6)$$

where $Expected[\overline{w}_c]$ is the expected number of broken machines in the system (either in queue or being serviced), given c repair crews. The hourly cost per disabled machine is difficult to estimate so we model it as a fuzzy number $\overline{B}$. We assume the hourly cost of each repair crew is known and crisp. The number of broken machines will be a fuzzy variable $\overline{w}_c$ having a discrete fuzzy probability distribution. For each value of $c = 1, 2, ..., N$ we find the fuzzy probability distribution of $\overline{w}_c$, its expected value and compute $\overline{Z}(c)$. Then we want the value of c to produce minimum $\overline{Z}(c)$.

Now there are three things to do: (1) discuss how we will find min $\overline{Z}(c)$, $1 \leq c \leq N$; (2) explain how we are to compute the discrete fuzzy probability distribution for $\overline{w}_c$; and (3) show how we are to obtain the expected value of $\overline{w}_c$ and finally get $\overline{Z}(c)$. Let us first consider minimizing fuzzy set $\overline{Z}(c)$. Now the $\overline{Z}(c)$ will be fuzzy numbers. So rank then, from smallest to largest, as discussed in Section 2.6. The ordering will partition the set of fuzzy numbers $\overline{Z}(c)$, $1 \leq c \leq N$, up into sets $H_1, ..., H_K$ with H_1 containing the smallest $\overline{Z}(c)$. Then the optimal values of c correspond to those $\overline{Z}(c)$ in H_1. If a unique optimal value of c is needed, then we will have to study more closely the relationship between the fuzzy numbers in H_1.

For each value of $c = 1, 2, ..., N$ we will have a fuzzy transition matrix $\overline{P}_c$ so that $\overline{P}_c^n \to \overline{\Pi}_c$ where each row in $\overline{\Pi}_c$ is $\overline{w}_c = (\overline{w}_{c0}, ..., \overline{w}_{cN})$. To illustrate the construction of these fuzzy transition matrices consider the following example.

Example 5.4.1.1

Let $N = 3$. Then we may have $c = 1, 2$ or $c = 3$. We will first discuss the crisp case. Let p be the probability of a machine failure in any time interval (assume the time interval is one hour). From past data we estimate $p \approx 0.03$. In the crisp case we would use $p = 0.03$. Next define $p(i, j)$ to be the probability that i machines fail in a one hour period, given that j machines had been broken down (and being serviced or waiting for a repair crew) at the end of the last time period, for $i = 0, 1, ..., 3 - j$ and $j = 0, 1, 2, 3$. We will treat $p(i, j)$ as a binomial probability

$$p(i, j) = \binom{3 - j}{i} p^i (1 - p)^{3 - j - i}, \tag{5.7}$$

for $i = 0, 1, ..., 3 - j$ and $j = 0, 1, 2, 3$.

Next define $q(i, j)$ to be the probability that i machines are repaired, and can go back into service, during a time period, given that there were j machines being repaired at the beginning of the time period, for $i = 0, 1, ..., j$ and $j = 1, 2, ..., c$. From past data we estimate that q, the probability that a repair crew will complete its job, which was started at the beginning of this period or during a previous time period, in this time period, is approximately 0.50. In the crisp case we would use $q = 0.50$. We assume that $q(i, j)$ may also be computed using the binomial probability function

$$q(i, j) = \binom{j}{i} q^i (1 - q)^{j - i}, \tag{5.8}$$

for $i = 0, 1, ..., j$ and $j = 0, 1, ..., c$.

Once we have determined all the $p(i, j)$ and $q(i, j)$ we can construct the transition matrices P_c, for $c = 1, 2, 3$. These matrices are shown in Tables 5.2-5.4.

Now we need to find the fuzzy transition matrices $\overline{P}_c$, $c = 1, 2, 3$. First we substitute $\overline{p} = (0.01/0.03/0.05)$ for p to show the uncertainty in the probability of a machine failure in a time interval. Using $\overline{p}$ we may determine the fuzzy probabilities $\overline{p}(i, j)$ from the fuzzy binomial in Chapter 4. Alpha-cuts of the $\overline{p}(i, j)$ are found as

$$\overline{p}(i, j)[\alpha] = \{ \binom{3 - j}{i} p^i (1 - p)^{3 - j - i} | p \in \overline{p}[\alpha] \}, \tag{5.9}$$

Table 5.2: The Transition Matrix P_1 in Example 5.4.1.

		Future State			
Repair crews	Previous state	0	1	2	3
c=1	0	$p(0,0)$	$p(1,0)$	$p(2,0)$	$p(3,0)$
	1	$p(0,1)q(1,1)$	$p(0,1)q(0,1)$ $+p(1,1)q(1,1)$	$p(1,1)q(0,1)$ $+p(2,1)q(1,1)$	$p(2,1)q(0,1)$
	2	0	$p(0,2)q(1,1)$	$p(0,2)q(0,1)$ $+p(1,2)q(1,1)$	$p(1,2)q(0,1)$
	3	0	0	$q(1,1)$	$q(0,1)$

Table 5.3: The Transition Matrix P_2 in Example 5.4.1.

		Future State			
Repair crews	Previous state	0	1	2	3
c=2	0	$p(0,0)$	$p(1,0)$	$p(2,0)$	$p(3,0)$
	1	$p(0,1)q(1,1)$	$p(0,1)q(0,1)$ $+p(1,1)q(1,1)$	$p(1,1)q(0,1)$ $+p(2,1)q(1,1)$	$p(2,1)q(0,1)$
	2	$p(0,2)q(2,2)$	$p(0,2)q(1,2)$ $+p(1,2)q(2,2)$	$p(0,2)q(0,2)$ $+p(1,2)q(1,2)$	$p(1,2)q(0,2)$
	3	0	$q(2,2)$	$q(1,2)$	$q(0,2)$

Table 5.4: The Transition Matrix P_3 in Example 5.4.1.

		Future State			
Repair crews	Previous state	0	1	2	3
c=3	0	$p(0,0)$	$p(1,0)$	$p(2,0)$	$p(3,0)$
	1	$p(0,1)q(1,1)$	$p(0,1)q(0,1)$ $+p(1,1)q(1,1)$	$p(1,1)q(0,1)$ $+p(2,1)q(1,1)$	$p(2,1)q(0,1)$
	2	$p(0,2)q(2,2)$	$p(0,2)q(1,2)$ $+p(1,2)q(2,2)$	$p(0,2)q(0,2)$ $+p(1,2)q(1,2)$	$p(1,2)q(0,2)$
	3	$q(3,3)$	$q(2,3)$	$q(1,3)$	$q(0,3)$

for all $\alpha \in [0,1]$. Most of these fuzzy probabilities are easy to find. For example, let us compute $\overline{p}(1,1)[\alpha] = [p(1,1)_1(\alpha), p(1,1)_2(\alpha)]$. Let $\overline{p}[\alpha] = [p_1(\alpha), p_2(\alpha)]$. Then

$$\overline{p}(1,1)[\alpha] = \{2p(1-p) | p \in \overline{p}[\alpha]\}. \tag{5.10}$$

But the function $f(p) = 2p(1-p)$ is increasing on the interval $\overline{p}[0] = [0.01, 0.05]$ so that $p(1,1)_1(\alpha) = 2p_1(\alpha)(1 - p_1(\alpha))$ and $p(1,1)_2(\alpha) = 2p_2(\alpha)(1 - p_2(\alpha))$.

Next we substitute $\overline{q} = (0.4/0.5/0.6)$ for q to describe the uncertainty in the probability q of a service completion during a time interval. Using $\overline{q}$ we may determine the fuzzy probabilities $\overline{q}(i,j)$ from the fuzzy binomial. Their α-cuts are

$$\overline{q}(i,j)[\alpha] = \{ \binom{j}{i} q^i (1-q)^{j-i} | q \in \overline{q}[\alpha] \}, \tag{5.11}$$

for $0 \le \alpha \le 1$. Most of these fuzzy probabilities are also easily computed.

Now we are ready to get the fuzzy transition matrices $\overline{P}_c$, $c = 1, 2, 3$. Let $\overline{P}_c = [\overline{p}_{cij}]$ a 4×4 matrix of fuzzy numbers. Let us illustrate how to get these fuzzy numbers $\overline{p}_{cij}$, $c = 1, 2, 3$ and $0 \le i, j \le 3$, by determining $\overline{p}_{112}$. The α-cuts of this fuzzy number are

$$\overline{p}_{112}[\alpha] = \{ p(1,1)q(0,1) + p(2,1)q(1,1) | \quad \mathbf{S} \quad \} \tag{5.12}$$

where $\mathbf{S}$ is the statement " $p(i,1) \in \overline{p}(i,1)[\alpha]$, $i = 0, 1, 2$, $\sum_{i=0}^{2} p(i,1) = 1$ and $q(i,1) \in \overline{q}(i,1)[\alpha]$, $i = 0, 1$, $q(0,1) + q(1,1) = 1$ ". All the fuzzy numbers in $\overline{P}_c$ are computed the same way for $c = 1, 2, 3$.

Once we have calculated the $\overline{P}_c$ we then need to find the $\overline{w}_c$. Then we need the fuzzy set $Expected[\overline{w}_c]$ whose α-cuts are

$$Expected[\overline{w}_c][\alpha] = \{ \sum_{i=0}^{N} i w_i | w_i \in \overline{w}_{ci}[\alpha], 0 \le i \le N, \sum_{i=0}^{N} w_i = 1 \}, \tag{5.13}$$

for $\alpha \in [0,1]$.

Finally, we multiply the two fuzzy numbers $\overline{B}$ and $Expected[\overline{w}_c]$ and add the crisp number Rc and we have calculated $\overline{Z}(c)$.

5.4.2 Fuzzy Queuing Decision Problem

In this queuing problem we have finite system (in the servers and in the queue) capacity M , c the number of identical and parallel servers, with finite calling source $N > M \ge c$ (total possible number of customers is N), and we wish to find the optimal number of servers c to minimize the total cost composed of the cost of the c servers plus the cost of lost customers. We want to minimize $\overline{Z}(c)$, a function of c, where

$$\overline{Z}(c) = C_1 c + \overline{C}_2 \{ Expected[\overline{L}_c] \}, \tag{5.14}$$

where C_1 is the cost per server in \$/unit time, $\overline{C}_2$ = cost (\$/unit time) per lost customer (turned away due to finite capacity), and $Expected[\overline{L}_c]$ is the expected number of lost customers. The constant $\overline{C}_2$ is a fuzzy number due to the fact that the price tag for lost customers is uncertain and hard to estimate. We assume that the constant C_1 is crisp and known. $\overline{L}_c$ is a fuzzy random variable having a discrete fuzzy distribution. For each value of $c = 1, 2, \ldots, M$ we find the fuzzy probability distribution of $\overline{L}_c$, its expected value and compute $\overline{Z}(c)$. Then we want the value of c to produce minimum $\overline{Z}(c)$.

We handle the problem of determining the minimum of $\overline{Z}(c)$, $1 \le c \le M$, as explained in the previous application.

To find the discrete fuzzy probability distribution of $\overline{L}_c$ we first use $M = N$. Given values for $\overline{p}$ and $\overline{q}$, we construct the fuzzy transition matrix $\overline{P}_c$, as described in Example 5.4.1.1, and find $\overline{P}_c^n \to \overline{\Pi}_c$ whose rows are $\overline{w}_c = (\overline{w}_{c0}, \ldots, \overline{w}_{cN})$. Now since we actually have finite capacity M the $\overline{w}_{ci}$, $i \ge M + 1$, give the fuzzy probabilities for lost customers. Then the α-cuts of $Expected[\overline{L}_c]$ are

$$Expected[\overline{L}_c][\alpha] = \{ \sum_{i=M+1}^{N} (i - M) w_i | w_i \in \overline{w}_{ci}[\alpha], 0 \le i \le N, \sum_{i=0}^{N} w_i = 1 \},$$

(5.15)

for all α. Multiply the fuzzy numbers $\overline{C}_2$ and $Expected[\overline{L}_c]$, add the crisp constant $C_1 c$, and we have computed $\overline{Z}(c)$

5.5 References

1. J.J.Buckley and E.Eslami: Uncertain Probabilities I: The Discrete Case, Soft Computing. To appear.

2. J.J.Buckley, T.Feuring and Y.Hayashi: Fuzzy Queuing Theory Revisited, Int. J. Uncertainty, Fuzziness and Knowledge-Based Systems, 9(2001), pp. 527-538.

Chapter 6

Fuzzy Markov Chains

6.1 Introduction

This Chapter continues our research into fuzzy Markov chains. In [4] we employed possibility distributions in finite Markov chains. The rows in a transition matrix were possibility distributions, instead of discrete probability distributions. Using possibilities we went on to look at regular, and absorbing, Markov chains and Markov decision processes.

The first three sections of this Chapter are based on [3]. We will show that the basic properties of regular, and absorbing, finite Markov chains carry over to our fuzzy Markov chains.

There have been a few other papers published on fuzzy Markov chains ([1],[2],[5],[7],[8],[9],[11]). In [8] the elements in the transition matrix are fuzzy probabilities, which are fuzzy subsets of $[0, 1]$, and we will do the same but under restricted fuzzy matrix multiplication (see below). That is, in [8] the authors use the extension principle to find powers of the fuzzy transition matrix which is different from our method which we call restricted fuzzy matrix multiplication. The paper [11] is more abstract and about a Markov fuzzy process with a transition possibility measure in an abstract state space. The paper [1] is like [4], containing good results about convergence of powers of the transition matrix. In [7] the authors use Dempster–Shafer type mass functions to construct transition probabilities for set–valued Markov chains in which the sets are subsets of the original state space. The authors in [2] were the first to consider stochastic systems in a fuzzy environment. By a "fuzzy environment" they mean the system has fuzzy goals and fuzzy constraints. Their transition matrix uses probabilities and they employed dynamic programming to obtain an optimal solution. This work was continued in [5] showing how fuzzy dynamic programming can be used to solve these types of problems. Fuzzy Markov decision problems were addressed in [9]. In this paper both the state and action are fuzzy, the transition of states is defined using a fuzzy relation, and the discounted total reward is described as a fuzzy number in a closed bounded interval. Our results are quite different from all of these other papers involving fuzzy Markov chains.

Let us now review some of the basic results from classical finite Markov chains ([6],[10]) (this material is a repeat from Section 5.2). A finite Markov chain has a finite number of possible states (outcomes) $S_1, S_2, \ldots, S_r$ at each step $n = 1, 2, 3\ldots$, in the process. Let

$$p_{ij} = Prob(S_j \ at \ step \ n + 1 | S_i \ at \ step \ n), \qquad (6.1)$$

$1 \leq i, j \leq r$, $n = 1, 2, \ldots$. The p_{ij} are the transition probabilities which do not depend on n. The transition matrix $P = (p_{ij})$ is a $r \times r$ matrix of the transition probabilities. An important property of P is that the row sums are equal to one and each $p_{ij} \geq 0$. Let $p_{ij}^{(n)}$ be the probability of starting in state S_i and ending up in S_j after n steps. Define P^n to be the product of P n–times and it is well known that $P^n = (p_{ij}^{(n)})$ for all n. If $p^{(0)} = (p_1^{(0)}, \ldots, p_r^{(0)})$, where $p_i^{(0)} =$ the probability of initially being in state S_i, and $p^{(n)} = (p_1^{(n)}, \cdots, p_r^{(n)})$, where $p_i^{(n)} =$ the probability of being in state S_i after n steps, we know that $p^{(n)} = p^{(0)} P^n$.

In the transition matrix $P = (p_{ij})$ all the p_{ij} must be known exactly. Many times these values are estimated or they are provided by "experts". We now assume that some of the p_{ij} are uncertain and we will model this uncertainty using fuzzy numbers. So, for each p_{ij} we substitute $\overline{p}_{ij}$ and define the fuzzy transition matrix $\overline{P} = (\overline{p}_{ij})$. Not all the p_{ij} need to be fuzzy, some can be crisp (a real number). If a p_{ij} is crisp we will still write it as $\overline{p}_{ij}$. If a $p_{ij} = 0$ or $p_{ij} = 1$, then we assume that there is no uncertainty in this value. If $0 < p_{ij} < 1$ and there is uncertainty in its value, then we assume that $0 < \overline{p}_{ij} < 1$ also.

The uncertainty is in some of the p_{ij} values but not in the fact that the rows in the transition matrix are discrete probability distributions. So we now put the following restriction on the $\overline{p}_{ij}$: there are $p_{ij} \in \overline{p}_{ij}[1]$ so that $P = (p_{ij})$ is the transition matrix for a finite Markov chain (the row sums equal one). This restriction on the p_{ij} is basic to the rest of the chapter.

Now we need to define restricted fuzzy matrix multiplication since we will need to compute $\overline{P}^n$ for $n = 2, 3, \ldots$ But first we require some definitions. Let

$$S = \{x = (x_1, \ldots, x_r) | x_i \geq 0, \sum_{i=1}^{r} x_i = 1\}, \qquad (6.2)$$

and then define

$$Dom_i[\alpha] = (\prod_{j=1}^{r} \overline{p}_{ij}[\alpha]) \bigcap S, \qquad (6.3)$$

for $0 \leq \alpha \leq 1$ and $1 \leq i \leq r$. Then (Dom for "domain")

$$Dom[\alpha] = \prod_{i=1}^{r} Dom_i[\alpha]. \qquad (6.4)$$

Next set $\overline{P}^n = (\overline{p}_{ij}^{(n)})$ where we will define $\overline{p}_{ij}^{(n)}$ and show that they are fuzzy numbers.

Consider a crisp transition matrix P and $P^n = (p_{ij}^{(n)})$. We know that

$$p_{ij}^{(n)} = f_{ij}^{(n)}(p_{11}, ..., p_{rr}), \qquad (6.5)$$

for some function $f_{ij}^{(n)}$. Equation (6.5) just says that the elements in P^n are some function of the elements in P. Now consider $f_{ij}^{(n)}$ a function of $p = (p_{11}, ..., p_{rr}) \in Dom[\alpha]$. Look at the range of $f_{ij}^{(n)}$ on $Dom[\alpha]$. Let

$$\Gamma_{ij}^{(n)}[\alpha] = f_{ij}^{(n)}(Dom[\alpha]). \qquad (6.6)$$

That is, $\Gamma_{ij}^{(n)}[\alpha]$ is the set of all values of $f_{ij}^{(n)}$ for $(p_{11}, ..., p_{rr}) \in Dom[\alpha]$. Now $f_{ij}^{(n)}$ is continuous and $Dom[\alpha]$ is connected, closed and bounded (compact), which implies that $\Gamma_{ij}^{(n)}[\alpha]$ is a closed and bounded interval for all α, i, j and n. We set

$$\overline{p}_{ij}^{(n)}[\alpha] = \Gamma_{ij}^{(n)}[\alpha], \qquad (6.7)$$

giving the α-cuts of the $\overline{p}_{ij}^{(n)}$ in $\overline{P}^n$. The resulting $\overline{p}_{ij}^{(n)}$ is a fuzzy number because its α-cuts are closed, bounded, intervals and surely it is normalized.

First, by restricting the $p_{ij} \in \overline{p}_{ij}[\alpha]$ to be in $Dom[\alpha]$, we get $P = (p_{ij})$ a crisp transition matrix for a finite Markov chain. Then an α-cut of $\overline{P}^n$ is the set of all P^n for $(p_{11}, ..., p_{rr}) \in Dom[\alpha]$. This is restricted fuzzy matrix multiplication because the uncertainties are in some of the p_{ij} values and not in the fact that each row in P must be a discrete probability distribution.

To compute the $\Gamma_{ij}^{(n)}[\alpha]$ all we need to find are the end points of the intervals. So we need to solve

$$p_{ij1}^{(n)}(\alpha) = min\{f_{ij}^{(n)}(p)|p \in Dom[\alpha]\}, \qquad (6.8)$$

and

$$p_{ij2}^{(n)}(\alpha) = max\{f_{ij}^{(n)}(p)|p \in Dom[\alpha]\}, \qquad (6.9)$$

where $\overline{p}_{ij}^{(n)}[\alpha] = [p_{ij1}^{(n)}(\alpha), p_{ij2}^{(n)}(\alpha)]$, all α.

In some simple cases, as shown in the examples in the next two sections, we can solve equations (6.8) and (6.9) for the α-cuts of the $\overline{p}_{ij}^{(n)}$. In general, one would need to employ a directed search algorithm (genetic, evolutionary) to estimate the solutions to equations (6.8) and (6.9).

It may appear that we are doing interval arithmetic to find the α-cuts in $\overline{P}^n$. Let us show that this is not the case. Let $\overline{P}[\alpha] = (\overline{p}_{ij}[\alpha])$ and $(\overline{P}[\alpha])^2 = (w_{ij}[\alpha])$ where

$$w_{ij}[\alpha] = \sum_{k=1}^{r} \overline{p}_{ik}[\alpha]\overline{p}_{kj}[\alpha], \tag{6.10}$$

all i, j and $\alpha \in [0, 1]$. Equation (6.10) is evaluated using interval arithmetic between all intervals. However, our restricted fuzzy matrix multiplication does not produce equation (6.10) for the α-cuts of $\overline{P}^2$. The following example, continued into the next section, shows the difference between the two methods and why using equation (6.10) is not useful in the study of finite fuzzy Markov chains.

Example 6.1.1

Let

$$P = \begin{pmatrix} 0.7 & 0.3 \\ 0.4 & 0.6 \end{pmatrix} \tag{6.11}$$

for a crisp transition matrix and then let $\overline{p}_{11} = (0.6/0.7/0.8)$, $\overline{p}_{12} = (0.2/0.3/0.4)$, $\overline{p}_{21} = (0.3/0.4/0.5)$ and $\overline{p}_{22} = (0.5/0.6/0.7)$. Use interval arithmetic, equation (6.10), to compute $\overline{P}^n$. $\overline{P}[\alpha] = (\overline{p}_{ij}[\alpha])$ where $\overline{p}_{11}[\alpha] = [0.6 + 0.1\alpha, 0.8 - 0.1\alpha]$, $\overline{p}_{12}[\alpha] = [0.2 + 0.1\alpha, 0.4 - 0.1\alpha]$, $\overline{p}_{21}[\alpha] = [0.3 + 0.1\alpha, 0.5 - 0.1\alpha]$ and $\overline{p}_{22}[\alpha] = [0.5 + 0.1\alpha, 0.7 - 0.1\alpha]$. Since all the intervals are non-negative we may find the end points of the intervals in $\overline{P}^n$ using

$$P_1(\alpha) = \begin{pmatrix} 0.6 + 0.1\alpha & 0.2 + 0.1\alpha \\ 0.3 + 0.1\alpha & 0.5 + 0.1\alpha \end{pmatrix}, \tag{6.12}$$

and

$$P_2(\alpha) = \begin{pmatrix} 0.8 - 0.1\alpha & 0.4 - 0.1\alpha \\ 0.5 - 0.1\alpha & 0.7 - 0.1\alpha \end{pmatrix}. \tag{6.13}$$

Let $\overline{P}^n = (w_{ij}^{(n)}(\alpha))$ where $w_{ij}^{(n)}[\alpha] = [w_{ij1}^{(n)}(\alpha), w_{ij2}^{(n)}(\alpha)]$. Then $P_1^n(\alpha) = (w_{ij1}^{(n)}(\alpha))$ and $P_2^n(\alpha) = (w_{ij2}^{(n)}(\alpha))$. But $w_{ij1}^{(n)}(\alpha) \to 0$ and $w_{ij2}^{(n)}(\alpha) \to \infty$ as $n \to \infty$ for $0 \le \alpha < 1$. In other words

$$w_{ij}^{(n)}(x) \to \begin{cases} 0, & x \le 0, \\ 1, & x > 1. \end{cases} \tag{6.14}$$

This is not a satisfactory result. We obtain better results using restricted fuzzy matrix multiplication in Example 6.2.1 in the next section.

6.2 Regular Markov Chains

We first review some of the basic properties of regular Markov chains. We say that the Markov chain is regular if $P^k > 0$ for some k, which is $p_{ij}^{(k)} > 0$ for all i, j. This means that it is possible to go from any state S_i to any state S_j in k steps. A property of regular Markov chains is that powers of P converge, or $\lim_{n \to \infty} P^n = \Pi$, where the rows of Π are identical. Let w be the unique left eigenvalue of P corresponding to eigenvalue one, so that $w_i > 0$ all i and $\sum_{i=1}^r w_i = 1$. That is $wP = w$ for $1 \times r$ vector w. Each row in Π is equal to w and $p^{(n)} \to p^{(0)}\Pi = w$. After a long time, thinking that each step being a time interval, the probability of being in state S_i is w_i, $1 \le i \le r$, independent of the initial conditions $p^{(0)}$. In a regular Markov chain the process goes on forever jumping from state to state, to state, ...

If P is a regular (crisp) Markov chain, then consider $\overline{P} = (\overline{p}_{ij})$ where $\overline{p}_{ij}$ gives the uncertainty (if any) in p_{ij}. If $(p_{11}, ..., p_{rr}) \in Dom[\alpha]$, then $P = (p_{ij})$ is also a regular Markov chain.

Let $\overline{P}^n \to \overline{\Pi}$ where each row in $\overline{\Pi}$ is $\overline{\pi} = (\overline{\pi}_1, ..., \overline{\pi}_n)$. Also let $\overline{\pi}_j[\alpha] = [\pi_{j1}(\alpha), \pi_{j2}(\alpha)]$, $1 \le j \le n$. We now show how to compute the α-cuts of the $\overline{\pi}_j$.

For each $(p_{11}, ..., p_{rr}) \in Dom[\alpha]$ set $P = (p_{ij})$ and we get $P^n \to \Pi$. Let $\Gamma(\alpha) = \{w | w \text{ } a \text{ } row \text{ } in \text{ } \Pi, (p_{11}, ..., p_{rr}) \in Dom[\alpha]\}$. $\Gamma(\alpha)$ consists of all vectors w, which are the rows in Π, for all $(p_{11}, ..., p_{rr}) \in Dom[\alpha]$. Then

$$\pi_{j1}(\alpha) = min\{w_j | w \in \Gamma(\alpha)\}, \tag{6.15}$$

and

$$\pi_{j2}(\alpha) = max\{w_j | w \in \Gamma(\alpha)\}. \tag{6.16}$$

In equations (6.15) and (6.16) w_j is the jth component in the vector w.

Example 6.2.1

This continues Example 6.1.1. If p is a 2×2 regular Markov chain, then we may find that $w_1 = p_{21}/(p_{21} + p_{12})$, $w_2 = p_{12}/(p_{21} + p_{12})$ where $w = (w_1, w_2)$ is a row in Π. Now we may solve equations (6.15) and (6.16) since $\partial w_1 / \partial p_{21} > 0$, $\partial w_1 / \partial p_{12} < 0$ and $\partial w_2 / \partial p_{21} < 0$, $\partial w_2 / \partial p_{12} > 0$. If $\overline{p}_{21}[\alpha] = [p_{211}(\alpha), p_{212}(\alpha)]$ and $\overline{p}_{12}[\alpha] = [p_{121}(\alpha), p_{122}(\alpha)]$ we obtain

$$\overline{\pi}_1[\alpha] = [\frac{p_{211}(\alpha)}{p_{211}(\alpha) + p_{122}(\alpha)}, \frac{p_{212}(\alpha)}{p_{212}(\alpha) + p_{121}(\alpha)}], \tag{6.17}$$

and

$$\overline{\pi}_2[\alpha] = [\frac{p_{121}(\alpha)}{p_{121}(\alpha) + p_{212}(\alpha)}, \frac{p_{122}(\alpha)}{p_{122}(\alpha) + p_{211}(\alpha)}], \tag{6.18}$$

both triangular fuzzy numbers. Equations (6.17) and (6.18) are correct because p_{12} and p_{21} are feasible (see Section 2.9). We know that, with restricted fuzzy matrix multiplication, $\overline{P}^n \to \overline{\Pi}$, where each row in $\overline{\Pi}$ is $(\overline{\pi}_1, \overline{\pi}_2)$. We may simplify equations (6.17) and (6.18) and get $\overline{\pi}_1 = (\frac{3}{7}/\frac{4}{7}/\frac{5}{7})$ and $\overline{\pi}_2 = (\frac{2}{7}/\frac{3}{7}/\frac{4}{7})$.

Example 6.2.2

Let

$$P = \begin{pmatrix} 0.5 & 0 & 0.5 \\ 0.25 & 0.75 & 0 \\ 0 & 0.6 & 0.4 \end{pmatrix} \tag{6.19}$$

be a transition matrix. $P^2 > 0$ so P is regular. If some p_{ij} are uncertain (substitute $\overline{p}_{ij}$), then another p_{ij} in the same row must also be uncertain since the row sums must equal one. Therefore, let $\overline{p}_{11} = \overline{p}_{13} = (0.4/0.5/0.6)$, $\overline{p}_{21} = (0.2/0.25/0.3)$, $\overline{p}_{22} = (0.7/0.75/0.8)$, $\overline{p}_{32} = (0.5/0.6/0.7)$, and $\overline{p}_{33} = (0.3/0.4/0.5)$. In this example we may also solve equations (6.15) and (6.16) because $w_1 = p_{32}p_{21}/S$, $w_2 = (1 - p_{11})p_{32}/S$, $w_3 = (1 - p_{11})p_{21}/S$ where $S = p_{32}p_{21} + (1-p_{11})p_{32} + (1-p_{11})p_{21}$. We determine that: (1) $\partial w_1/\partial p_{11} > 0$, $\partial w_1/\partial p_{21} > 0$, $\partial w_1/\partial p_{32} > 0$; (2) $\partial w_2/\partial p_{11} < 0, \partial w_2/\partial p_{21} < 0$, $\partial w_2/\partial p_{32} > 0$; and (3) $\partial w_3/\partial p_{11} < 0$, $\partial w_3/\partial p_{21} > 0$, $\partial w_3/\partial p_{32} < 0$. This allows us to find the α-cuts of the $\overline{\pi}_i$, $i = 1, 2, 3$. The solution for the end points of the α-cuts is not difficult because the p'_is are feasible (Section 2.9). Let us illustrate this with $\overline{\pi}_2[\alpha]$. Let $\overline{\pi}_2[\alpha] = [\pi_{21}(\alpha), \pi_{22}(\alpha)]$ and let $\overline{p}_{ij}[\alpha] = [p_{ij1}(\alpha), p_{ij2}(\alpha)]$ all i, j. From the above analysis we see that

$$\pi_{21}(\alpha) = (1 - p_{112}(\alpha))p_{321}(\alpha)/S_1(\alpha), \tag{6.20}$$

where

$$S_1(\alpha) = p_{321}(\alpha)p_{212}(\alpha) + (1-p_{112}(\alpha))p_{321}(\alpha) + (1-p_{112}(\alpha))p_{212}(\alpha), \tag{6.21}$$

for $\alpha \in [0, 1]$. We can use the end points of the α-cuts of $\overline{p}_{11}$, $\overline{p}_{21}$ and $\overline{p}_{32}$ because p_{11}, p_{21}, p_{32} are feasible. Also

$$\pi_{22}(\alpha) = (1 - p_{111}(\alpha))p_{322}(\alpha)/S_2(\alpha), \tag{6.22}$$

where

$$S_2(\alpha) = p_{322}(\alpha)p_{211}(\alpha) + (1-p_{111}(\alpha))p_{322}(\alpha) + (1-p_{111}(\alpha))p_{211}(\alpha), \tag{6.23}$$

for all α.

The $\overline{\pi}_i$ will be triangular shaped fuzzy numbers whose $\alpha = 0$ and $\alpha = 1$ cuts are presented in Table 6.1. Triangular "shaped" means that the sides of the "triangle" are curves not straight lines. Then $\overline{P}^n \to \overline{\Pi}$ where the rows of $\overline{\Pi}$ are $(\overline{\pi}_1, \overline{\pi}_2, \overline{\pi}_3)$.

	$\alpha = 1$	$\alpha = 0$
$\overline{\pi}_1$	0.2609	[0.1923,0.3443]
$\overline{\pi}_2$	0.5217	[0.4255,0.6176]
$\overline{\pi}_3$	0.2174	[0.1600,0.2857]

Table 6.1: Alpha-cuts of the Fuzzy Numbers $\overline{\pi}_i$ in Example 6.2.2.

6.3 Absorbing Markov Chains

First we will discuss the basic results for crisp absorbing Markov chains. We will call a state S_i absorbing if $p_{ii} = 1$, $p_{ij} = 0$ for $i \neq j$. Once in S_i you can never leave. Suppose there are k absorbing states, $1 \leq k < r$, and then we may rename the states (if needed) so that the transition matrix P can be written as

$$P = \begin{pmatrix} I & O \\ R & Q \end{pmatrix}, \qquad (6.24)$$

where I is the $k \times k$ identity, O is the $k \times (r-k)$ zero matrix, R is $(r-k) \times k$ and Q is $(r-k) \times (r-k)$. The Markov chain is called an absorbing Markov chain if it has at least one absorbing state and from every non–absorbing state it is possible to reach some absorbing state in a finite number of steps. Assume the chain is absorbing and then we know that

$$P^n = \begin{pmatrix} I & O \\ SR & Q^n \end{pmatrix}, \qquad (6.25)$$

where $S = I + Q + \cdots + Q^{n-1}$. Then $\lim_{n \to \infty} P^n = \Pi$ where

$$\Pi = \begin{pmatrix} I & O \\ R^* & O \end{pmatrix}, \qquad (6.26)$$

for $R^* = (I - Q)^{-1}R$. Notice the zero columns in Π which implies that the probability that the process will eventually enter an absorbing state is one. The process eventually ends up in an absorbing state.

If $R = (r_{ij})$ and $Q = (q_{ij})$ we now assume that there is uncertainty in some of the r_{ij} and/or the q_{ij} values. We then substitute $\overline{r}_{ij}$ for r_{ij} and $\overline{q}_{ij}$ for q_{ij} and obtain $\overline{P}$ an absorbing fuzzy Markov chain. We now show, under restrictive fuzzy matrix multiplication, that $\overline{P}^n \to \overline{\Pi}$ where

$$\overline{\Pi} = \begin{pmatrix} I & O \\ \overline{R}^* & O \end{pmatrix} \qquad (6.27)$$

with $(r-k) \times k$ matrix $\overline{R}^* = (\overline{r}_{ij}^*)$. For any $p \in Dom[\alpha]$, P^n converges to the Π in equation (6.26) which implies that $\overline{Q}^n \to O$, the (crisp) zero matrix. Also,

for any $p \in Dom[\alpha]$, $R^* = (I - Q)^{-1}R = (r_{ij}^*)$. Let $\bar{r}_{ij}^*[\alpha] = [r_{ij1}^*(\alpha), r_{ij2}^*(\alpha)]$.
It follows that

$$r_{ij1}^*(\alpha) = min\{r_{ij}^* | p \in Dom[\alpha]\}, \tag{6.28}$$

and

$$r_{ij2}^*(\alpha) = max\{r_{ij}^* | p \in Dom[\alpha]\}. \tag{6.29}$$

To find the limit of $\overline{P}^n$, as $n \to \infty$, which is $\overline{\Pi}$, all we need to do is solve
equations (6.28) and (6.29) for the α-cuts of the $\bar{r}_{ij}^*$ in $\overline{R}^*$ in equation (6.27).

Example 6.3.1

Let

$$P = \begin{pmatrix} 1 & 0 & 0 \\ 0.2 & 0.6 & 0.2 \\ 0.4 & 0.3 & 0.3 \end{pmatrix} \tag{6.30}$$

have one absorbing state. Substitute $\bar{p}_{ij}$ for p_{ij} expressing the uncertainty in
the p_{ij} values. Then $\overline{P}^n \to \overline{\Pi}$ where $\overline{\Pi}$ is the crisp matrix

$$\overline{\Pi} = \begin{pmatrix} 1 & 0 & 0 \\ 1 & 0 & 0 \\ 1 & 0 & 0 \end{pmatrix}, \tag{6.31}$$

because $\overline{Q}^n \to O$ and the $\bar{r}_{i1}^*$ must equal crisp one because the row sums are
one.

Example 6.3.2

We next consider two absorbing states with transition matrix

$$P = \begin{pmatrix} 1 & 0 & 0 & 0 \\ 0 & 1 & 0 & 0 \\ 0.4 & 0 & 0 & 0.6 \\ 0 & 0.6 & 0.4 & 0 \end{pmatrix}. \tag{6.32}$$

Substitute $\bar{r}_{ij}$ for r_{ij} and $\bar{q}_{ij}$ for q_{ij} where $\bar{r}_{11} = (0.3/0.4/0.5)$, $\bar{r}_{22} = (0.5/0.6/0.7)$, $\bar{q}_{12} = (0.5/0.6/0.7)$ and $\bar{q}_{21} = (0.3/0.4/0.5)$. First we determine the r_{ij}^* values in terms of the r_{ij} and the q_{ij}. Using $r_{11} + q_{12} = 1$
and $r_{22} + q_{21} = 1$ we obtain $r_{11}^* = r_{11}/T_1$, $r_{12}^* = r_{22}q_{12}/T_2$, $r_{21}^* = r_{11}q_{21}/T_1$
and $r_{22}^* = r_{22}/T_2$ where $T_1 = 1 - q_{21} + r_{11}q_{21}$ and $T_2 = 1 - q_{12} + r_{22}q_{12}$.

We may solve equations (6.28) and (6.29) by noting that each r_{ij}^* is an
increasing function of all its arguments. For example, this means that r_{21}^* is
an increasing function of r_{11} and q_{21}. Let $\bar{r}_{21}^*[\alpha] = [L_{21}(\alpha), R_{21}(\alpha)]$, $\bar{r}_{ij}[\alpha] = [r_{ij1}(\alpha), r_{ij2}(\alpha)]$ and $\bar{q}_{ij}[\alpha] = [q_{ij1}(\alpha), q_{ij2}(\alpha)]$. Then

$$L_{21}(\alpha) = r_{111}(\alpha)q_{211}(\alpha)/T_{11}(\alpha), \tag{6.33}$$

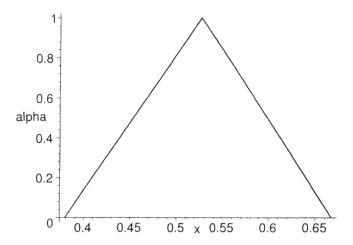

Figure 6.1: Fuzzy Number $\bar{r}_{11}^*$ in Example 6.3.2

where
$$T_{11}(\alpha) = 1 - q_{211}(\alpha) + r_{111}(\alpha)q_{211}(\alpha), \tag{6.34}$$

since r_{11} and q_{21} are feasible. Also

$$R_{21}(\alpha) = r_{112}(\alpha)q_{212}(\alpha)/T_{12}(\alpha), \tag{6.35}$$

where
$$T_{12}(\alpha) = 1 - q_{212}(\alpha) + r_{112}(\alpha)q_{212}(\alpha). \tag{6.36}$$

The graphs of the $\bar{r}_{ij}^*$ are in Figures 6.1 through 6.4.

6.4   Application: Decision Model

This application may be found in [4], however there we employed possibilities and here we use fuzzy probabilities. We will be working with a finite horizon Markov decision model having K steps. At each step the decision maker has two choices: (1) make decision D_a; or (2) choose decision D_b. We consider only two decisions but can easily generalize to any finite number of decisions at each step. If the decision is D_a (D_b), then the fuzzy transition matrix is $\overline{P}_a = (\overline{p}_{aij})$ ($\overline{P}_b = (\overline{p}_{bij})$), a $r \times r$ matrix of fuzzy transition probabilities from state S_i to state S_j. Also, after making the decision D_a (D_b) the decision maker receives a gain, or loss, represented by $r \times r$ matrix $R_a = (r_{aij})$ ($R_b = (r_{bij})$). For example, if at step k we observe the system and it is in state S_i and the decision maker chooses D_b then: (1) we use the i-th row in $\overline{P}_b$ for the fuzzy transition possibilities to the next state; and (2) we use the

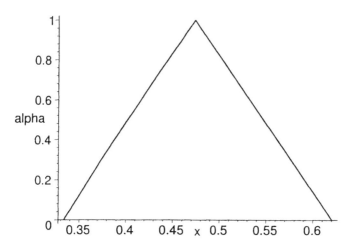

Figure 6.2: Fuzzy Number $\bar{r}_{12}^*$ in Example 6.3.2

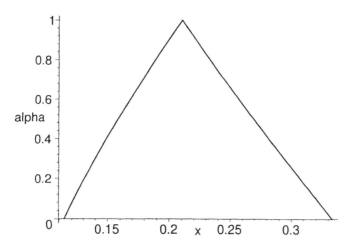

Figure 6.3: Fuzzy Number $\bar{r}_{21}^*$ in Example 6.3.2

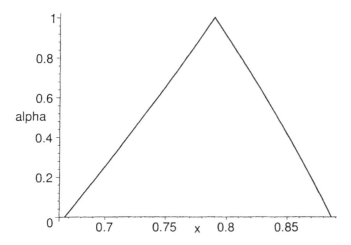

Figure 6.4: Fuzzy Number $\bar{r}_{22}^*$ in Example 6.3.2

i-th row in R_b to compute the next gain, or loss, corresponding to the next state. If in addition the next state is S_j then the fuzzy probability of the transition is $\bar{p}_{bij}$ and the gain/loss is r_{bij}. The elements in R_a and R_b are all in dollars.

The policy space Γ consists of all sequences of D_a, or D_b, of length K. We rename these sequences in Γ by A_i, $1 \leq i \leq M$. For example, if $K = 3$, then $A_1 = D_a D_a D_a$, $A_2 = D_a D_a D_b$, ..., $A_8 = D_b D_b D_b$. In general, $M = 2^K$. A policy will then be an $A_i \in \Gamma$.

We wish to find the "best" policy in Γ. Given $\overline{P}_a$ $(\overline{P}_b)$ and R_a (R_b) we want to find the $A_i \in \Gamma$ that, in some sense, maximizes the return to the decision maker. In crisp probability based Markov chains one usually finds the policy that will maximize the expected return over the finite horizon (K steps) using dynamic programming. We propose to do the same thing, without fuzzy dynamic programming, using fuzzy probabilities.

For each initial state S_i we want the optimal policy. That is, we will not be using an initial fuzzy possibility distribution $\bar{p}^{(0)}$ and instead we assume we will always know in what state the process will begin. Suppose we start in state S_j, then once we choose a policy A_i in Γ it determines a discrete fuzzy probability distribution $\overline{Prob}(A_i)$ over the returns (losses) incurred by the decision maker. The following example shows how we will compute $\overline{Prob}(A_i)$, all A_i in Γ.

Example 6.4.1

Suppose $K = 3$ so that Γ has 8 members. Also assume that $r = 3$ so that there is only three states. Let us find $\overline{Prob}(A_3)$ for $A_3 = D_a D_b D_a$ and assume we start in state S_2. Let

$$\tau_{ijk} = 2 D_a i D_b j D_a k \ , \tag{6.37}$$

$1 \leq i, j, k \leq 3$, which represents a possible outcome. The expression for τ_{ijk} is interpreted as follows: (1) we start in state S_2, choose D_a, end up in state S_i; (2) at the start of step 2 we are in state S_i, choose D_b, and go to S_j; and (3) at the start of step 3 we are in S_j, choose decision D_a, and change to state S_k. The return (loss) r_{ijk} to the decision maker corresponding to a τ_{ijk} is

$$r_{ijk} = \delta r_{a2i} + \delta^2 r_{bij} + \delta^3 r_{ajk} \ , \tag{6.38}$$

discounting future monies back to their present value using $\delta = (1 + \beta)^{-1}$ for interest rate β. If the steps represent very short time periods, then we would not discount and set $\delta = 1$. The fuzzy probability of this outcome is $\overline{p}_{ijk}$ whose α-cuts are

$$\overline{p}_{ijk}[\alpha] = \{ p_{a2i} p_{bij} p_{ajk} | \quad \mathbf{S} \quad \}, \tag{6.39}$$

for all α where $\mathbf{S}$ denotes the statement "$p_{xyz} \in \overline{p}_{xyz}[\alpha]$, $x = a, b$, $1 \leq y, z \leq 3$ and the row sums in the crisp transition matrices $P_a = (p_{aij})$ and $P_b = (p_{bij})$ all equal one". This produces a correct discrete fuzzy probability distribution over the values of the r_{ijk} because it is not difficult to see that for any α one can choose $p_{ijk} \in \overline{p}_{ijk}[\alpha]$ whose sum is one. There is a possibility that two (or more) τ_{ijk} values might be the same. Then we add the corresponding fuzzy probabilities to get the fuzzy probability of their common value.

Now we can present our solution. Assume the same situation as in Example 6.4.1 with $K = 3$, $r = 3$, etc. Fix the initial state and compute $\overline{Prob}(A_i)$ for all A_i. Then we find α-cuts of the expected value of $\overline{Prob}(A_i)$ as

$$Expected[\overline{Prob}(A_i)][\alpha] = \{ \sum_{i,j,k} r_{ijk} p_{ijk} | \quad \mathbf{S} \quad \}, \tag{6.40}$$

for $0 \leq \alpha \leq 1$ and $\mathbf{S}$ is "$p_{ijk} \in \overline{p}_{ijk}[\alpha]$ all i, j, k and the row sums in the crisp transition matrices are all equal to one". Then we rank the fuzzy numbers $Expected[\overline{Prob}(A_i)]$ from smallest to largest as discussed in Section 2.6. This gives a set H_K of highest ranked fuzzy numbers. Policies A_i corresponding to those $Expected[\overline{Prob}(A_i)]$ in H_K are then the optimal policies for this initial state.

6.5 References

1. K.E.Avrachenkov and E.Sanchez: Fuzzy Markov Chains: Specificities and Properties, Proc. IPMU, July 3-7, Madrid, Spain, 2000, pp. 1851-1856.

2. R.E.Bellman and L.A.Zadel: Decision-Making in a Fuzzy Environment, Management Science, 17(1970), pp. 141-164.

3. J.J.Buckley and E.Eslami: Fuzzy Markov Chains: Uncertain Probabilities, Mathware and Soft Computing. To appear.

4. J.J.Buckley, T.Feuring and Y.Hayashi: Fuzzy Markov Chains,under review.

5. A.O.Esogbue and R.E.Bellman: Fuzzy Dynamic Programming and its Extensions, TIMS Studies in the Management Sciences, 20(1984), pp. 147-167.

6. J.G.Kemeny and J.L.Snell: Finite Markov Chains, Van Nostrand, N.Y., 1960.

7. R.M. Kleyle and A.deKorvin: Constructing One-Step and Limiting Fuzzy Transition Probabilities for Finite Markov Chains, J. Intelligent and Fuzzy Systems, 6(1998), pp. 223-235.

8. R.Kruse, R.Buck-Emden and R.Cordes, Processor Power Considerations - An Application of Fuzzy Markov Chains, Fuzzy Sets and Systems, 21(1987), pp. 289-299.

9. M.Kurano, M.Yasuda, J.Nakagami and Y.Yoshida, Markov-type Fuzzy Decision Processes with a Discounted Reward on a Closed Interval, European J. Oper. Res., 92(1996), pp. 649-662.

10. M.Olinick, An Introduction to Mathematical Models in the Social and Life Sciences, Addison-Wesley , Reading, Mass., 1978.

11. Y.Yoshida, Markov Chains with a Transition Possibility Measure and Fuzzy Dynamic Programming, Fuzzy Sets and Systems, 66(1994), pp. 39-57.

Chapter 7

Fuzzy Decisions Under Risk

7.1   Introduction

A decision problem under risk is characterized by three things: (1) the decision maker must choose an action a_i from a finite set of actions $\mathcal{A} = \{a_1, ..., a_m\}$; (2) a finite set of chance events (also called the states of nature) $\mathcal{E} = \{e_1, ..., e_n\}$ over which we have no control; and (3) for each a_i and resulting e_j a payoff $c(a_i, e_j)$ which describes what happens when event e_j occurs after the choice of action a_i. We assume that the payoff $c(a_i, e_j)$ is measured in dollars and the decision maker wants to maximize his payoff. Sometimes one will use a utility function for the decision maker for the payoff, but we shall not discuss utility theory in this book. What makes this decision making under risk is that we now assume that there is a probability distribution over $\mathcal{E}$ giving the probability of each event e_j. Let the probability of e_j be p_j, $1 \leq j \leq n$. So the whole decision problem may be described by a $m \times n$ matrix M where; (1) the rows are labeled by the actions $a_1, ..., a_m$; (2) the columns are labeled by the events $e_1, ..., e_n$; (3) the ij^{th} element in M is the payoff $c(a_i, e_j)$; and (4) the probabilities p_j are placed over the events e_j.

The probabilities p_j are usually unknown and must be estimated, or obtained from experts (see Section 2.7). Therefore, we will substitute fuzzy numbers $\bar{p}_j$ for the p_j giving a fuzzy decision making problem under risk. We will study this problem (called the "without data problem") in the next section followed in Section 7.3 by the "with data" problem where we first obtain some data on the possible events e_j. In Section 7.2 we will look at two possible solution strategies: (1) maximize expected payoff ; and (2) satisfy some aspiration level. In Section 7.3 we only apply the maximize expected payoff method. In both sections we will illustrate the results only through worked examples.

		0.6	0.4
-------		------	------
		e_1	e_2
a_1		2	5
a_2		4	1.5
a_3		6	0

Table 7.1: Decision Problem in Example 7.2.1

Decision making under risk problems are discussed in most operations research/management science books, and also in game theory books and some statistics books. Let us only give a few key references ([1],[2],[3],[4]).

7.2   Without Data

Our first example uses the solution method of maximizing expected payoff. We first work through the crisp case followed by the fuzzy probabilities.

Example 7.2.1

The decision problem is shown in Table 7.1. To find the expected payoff we compute $E(a_1) = (0.6)(2)+(0.4)(5) = 3.2$, $E(a_2) = (0.6)(4)+(0.4)(1.5) = 3.0$ and $E(a_3) = (0.6)(6) + (0.4)(0) = 3.6$. Hence, the maximum is 3.6 and the optimal action is a_3.

Now we go to the fuzzy decision problem. We will substitute $\bar{p}_1 = (0.5/0.6/0.7)$ for p_1 and $\bar{p}_2 = (0.3/0.4/0.5)$ for p_2. Let $\bar{A}_i$ be the fuzzy expected payoff given action a_i was chosen, $1 \leq i \leq 3$. Then the α-cuts of these fuzzy probabilities are

$$\bar{A}_i[\alpha] = \{\sum_{j=1}^{2} c(a_i, e_j)p_j \mid p_j \in \bar{p}_j[\alpha], p_1 + p_2 = 1\}, \qquad (7.1)$$

for all α in $[0, 1]$. Now rank the $\bar{A}_i$, from smallest to largest, as described in Section 2.6, and choose those actions that correspond to the highest ranked $\bar{A}_i$ as the optimal actions.

In this simple example it is easy to compute the $\bar{A}_i$, since we may use $p_2 = 1 - p_1$ in equation (7.1). The result is: (1) $\bar{A}_1 = (2.9/3.2/3.5)$; (2) $\bar{A}_2 = (1.25/1.5/1.75)$; and (3) $\bar{A}_3 = (3/3.6/4.2)$. Then the highest ranked fuzzy set is $\bar{A}_3$, because $\bar{A}_1 < \bar{A}_3$ using $\eta = 0.8$ in Section 2.6, and the optimal action continues to be a_3.

The second example uses the aspiration level method of solution.

I	ES	EE
0	2.3	0
1	1.4	0.1
2	0.7	0.4
3	0.3	1.0
4	0.1	1.8
5	0	2.7

Table 7.2: Crisp Solution in Example 7.2.2

Example 7.2.2

This example was adapted from an example in ([4], p.418). Suppose we have a commodity whose demand per period can take on values $0, 1, 2, 3, 4, 5$ with probability $0.1, 0.2, 0.3, 0.2, 0.1, 0.1$, respectively. Let I be the inventory at the start of the period and I can take on the values $0, 1, 2, 3, 4, 5$. So the actions by the decision maker are the values of I and the events are the values of demand for the period. In our previous notation $a_i \in \{0, 1, 2, 3, 4, 5\}$ and $e_j \in \{0, 1, 2, 3, 4, 5\}$ with associated probabilities p_j given above.

If the amount in inventory at the start of the period is not sufficient, shortage may occur with loss of profit and loss of customer goodwill. The expected shortage (ES) is

$$ES = \sum_{j=I+1}^{5} (j - I)p_j, \tag{7.2}$$

if $I < 5$, and it equals zero for $I = 5$. If we start the period with too much inventory, then excess inventory may occur with an increase in the cost of storing and maintaining this commodity. The expected excess (EE) is

$$EE = \sum_{j=0}^{I-1} (I - j)p_j, \tag{7.3}$$

if $I > 0$, and it is zero when $I = 0$.

The decision maker wants to balance these two conflicting costs. This decision maker decides on two numbers A_1 and A_2 and sets the goals: (1) $ES \leq A_1$; and (2) $EE \leq A_2$. This is the aspiration model. We do not try to maximize (or minimize) anything. We set, in the cost case, maximum levels we wish to avoid and obtain values less than. Suppose $A_1 = A_2 = 1$. Then the solution is shown in Table 7.2. We see that $ES \leq 1$ implies that $I \geq 2$ and $EE \leq 1$ is true when $I \leq 3$. Hence, the optimal solution is $I = 2, 3$.

Assume that the probability that demand equals $0, 1, ..., 5$ has to be estimated, or given by experts. Then we construct and use fuzzy numbers for these probabilities. Let $\bar{p}_0 = (0.05/0.10/0.15)$, $\bar{p}_1 = (0.1/0.2/0.3)$, $\bar{p}_2 = (0.2/0.3/0.4)$, $\bar{p}_3 = (0.1/0.2/0.3)$, $\bar{p}_4 = (0.05/0.10/0.15)$, and $\bar{p}_5 = (0.05/0.10/0.15)$. Then EE and ES become fuzzy numbers. For example, α-cuts of $\overline{ES}$ are

$$\overline{ES}[\alpha] = \{ \sum_{j=I+1}^{5} (j-I)p_j | p_j \in \bar{p}_j[\alpha], \sum_{j=0}^{5} p_j = 1 \}, \qquad (7.4)$$

if $I < 5$. However, $\overline{ES}$ and $\overline{EE}$ turnout to be triangular fuzzy numbers or triangular shaped fuzzy numbers, when they are not zero. The results are displayed in Table 7.3. The notation $\approx (a/b/c)$ means it is possibly a triangular shaped fuzzy number with support $[a, c]$ and vertex at $x = b$. We used the numerical method "simplex" discussed in Section 2.9 to find the supports of the triangular shaped fuzzy numbers. In Table 7.3 whenever there is a $\approx$ we had to use the "simplex" method because the $p_i's$ were not feasible, in the other cases the $p_i's$ were feasible (see Section 2.9). For example, the optimization problem to solve for the support of $\overline{ES}$ when $I = 0$ is

$$max/min\{p_1 + 2p_2 + 3p_3 + 4p_4 + 5p_5\}, \qquad (7.5)$$

subject to

$$p_i \in \bar{p}_i[0], \sum_{i=0}^{5} p_i = 1. \qquad (7.6)$$

All we need now are fuzzy numbers for A_1 and A_2. Let $\overline{A}_1 = \overline{A}_2 = (0/1/2)$. Now we need to decide on the values of I for which $\overline{ES} \leq \overline{A}_1$ and also those I values so that $\overline{EE} \leq \overline{A}_2$. For these comparisons we employ the method given in Section 2.6 and, for simplicity, we used triangular fuzzy numbers, and not triangular shaped fuzzy numbers, in all the comparisons. In this example let us use $\eta = 0.7$. Then we easily see that $\overline{ES} \leq \overline{A}_1$ for $I \geq 2$ and $\overline{EE} \leq \overline{A}_2$ for $I \leq 3$. Now for $I = 1$ we compute $v(\overline{ES} \leq \overline{A}_1) = 0.73$ and $v(\overline{A}_1 \leq \overline{ES}) = 1$. Hence for $\eta = 0.7$ and $I = 1$ we have $\overline{ES} \approx \overline{A}_1$ and $\overline{ES} \leq \overline{A}_1$. Now for $I = 4$, we determine that $\overline{A}_2 < \overline{EE}$. Hence, in the fuzzy case, the optimal values of I are $1, 2, 3$.

7.3   With Data

Again, we first present the crisp problem and then the fuzzy case using the plan of choosing the action that will maximize the expected payoff.

Example 7.3.1

This example, adapted from examples in ([2], Chapter 8), continues Example 7.2.1, however we now gather some data (information) about which chance

I	$\overline{ES}$	$\overline{EE}$
0	$\approx (1.75/2.3/2.85)$	0
1	$\approx (0.90/1.4/1.90)$	$(0.05/0.10/0.15)$
2	$(0.35/0.7/1.05)$	$(0.2/0.4/0.6)$
3	$(0.15/0.3/0.45)$	$\approx (0.60/1/1.40)$
4	$(0.05/0.1/0.15)$	$\approx (1.3/1.8/2.3)$
5	0	$\approx (2.15/2.7/3.25)$

Table 7.3: Fuzzy Expected Values in Example 7.2.2

	$f(z|e_1)$	$f(z|e_2)$	$f(z)$
z_1	0.7	0.1	0.46
z_2	0.3	0.9	0.54
$g(e)$	0.6	0.4	

Table 7.4: Conditional Probabilities in Example 7.3.1

event might occur before we choose our action. Suppose e_1 is the event that it will rain tomorrow and e_2 is the event that there will be no rain tomorrow. A single observation of a rain indicator (or a weather report) will be our data Z. The random variable Z will have two values z_1 and z_2. The value z_1 is that the indicator predicts rain tomorrow and z_2 predicts no rain tomorrow. The conditional probabilities $f(z_k|e_j)$ are given in Table 7.4. For example, assume from past experience that when it is going to rain tomorrow (event e_1), the probability is 0.7 that the indicator will show rain ($Z = z_1$). Past experience (data, information) is used to compute the other probabilities in Table 7.4. The original probability distribution over the chance events is now called the prior distribution $g(e)$ also shown in Table 7.4. We used $g(e_1) = 0.6$ and $g(e_2) = 0.4$ in Example 7.2.1. We will explain the probability column $f(z)$ below.

We find the joint probability mass function (see also Chapter 10) as

$$f(z_k, e_j) = f(z_k|e_j)g(e_j). \tag{7.7}$$

From the joint we calculate the marginal $f(z_k)$, shown in Table 7.4, as follows

$$f(z_k) = \sum_{j=1}^{2} f(z_k, e_j). \tag{7.8}$$

From this we may find the new probability distribution over the chance events, now called the posterior distribution, from

$$f(e_j|z_k) = f(z_k, e_j)/f(z_k), \tag{7.9}$$

	$f(e\|z_1)$	$f(e\|z_2)$
e_1	42/46	18/54
e_2	4/46	36/54

Table 7.5: Posterior Probabilities in Example 7.3.1

	z_1	z_2
a_1	2.26	4.00
a_2	3.78	2.33
a_3	5.48	2.00

Table 7.6: Final Expected Payoff in Example 7.3.1

which was calculated and is given in Table 7.5.

Using Tables 7.1 and 7.5 we find the new expected payoffs. For each value of Z the $f(e_j|z)$ becomes the new probability mass function over the chance events. The calculation for the expected payoffs is done as follows

$$E[a_i|z_k] = \sum_{j=1}^{2} c(a_i, e_j)f(e_j|z_k). \qquad (7.10)$$

Then for each z_k value the optimal action is the one that maximizes this expected value. The expected values are displayed in Table 7.6. We see if $Z = z_1$, then the best action is a_3 and when $Z = z_2$ we would use a_1.

The probabilities in Table 7.4 have all been estimated from past data so, as discussed in Section 2.7, we will substitute fuzzy numbers for these probabilities. Let $q_{kj} = f(z_k|e_j)$ and then set $\bar{q}_{11} = (0.5/0.7/0.9)$, $\bar{q}_{21} = (0.1/0.3/0.5)$, $\bar{q}_{12} = (0/0.1/0.2)$ and $\bar{q}_{22} = (0.8/0.9/1)$. We will use the fuzzy probabilities $\bar{p}_j$ from Example 7.2.1. So, $\bar{g}(e_1) = \bar{p}_1 = (0.5/0.6/0.7)$ and $\bar{g}(e_2) = \bar{p}_2 = (0.3/0.4/0.5)$. What we need to do first is to calculate the new posterior fuzzy probabilities $\bar{w}_{jk} = \bar{f}(e_j|z_k)$. We will determine their α-cuts as

$$\bar{w}_{jk}[\alpha] = \{\frac{q_{kj}p_j}{\sum_{j=1}^{2} q_{kj}p_j}| \text{ S } \}, \qquad (7.11)$$

where S is the statement "$p_i \in \bar{p}_i[\alpha], p_1 + p_2 = 1, q_{kj} \in \bar{q}_{kj}[\alpha], \sum_{k=1}^{2} q_{kj} = 1$, all j", all $\alpha \in [0, 1]$. An alternate method of calculating these fuzzy probabilities is to first determine the fuzzy numbers for the joint $\bar{f}(z_k, e_j)$ and the marginal $\bar{f}(z_k)$ and then divide these two fuzzy numbers. However, just like in Section 3.3 on fuzzy conditional probabilities, we reject this alternate procedure in favor of equation (7.11). Now let $\bar{A}_{ik} = \bar{E}[a_i|z_k]$ be the fuzzy expected payoff of action a_i given $Z = z_k$ using the posterior fuzzy probabilities. Then we find α-cuts of these fuzzy numbers as

$$\overline{A}_{ik}[\alpha] = \{\sum_{j=1}^{2} c(a_i, e_j) w_{jk} | w_{jk} \in \overline{w}_{jk}[\alpha], \sum_{j=1}^{2} w_{jk} = 1\}. \tag{7.12}$$

Then, for each value of Z, the optimal action corresponds to the largest values of $\overline{A}_{ik}$ as discussed in Section 2.6.

For a value of Z we now show how we may find the $\overline{A}_{ik}$, by first determining their α-cuts, and hence solve the problem. Let $Z = z_1$ and then we calculate $\overline{A}_{i1}$ for $i = 1, 2, 3$. First we need the $\overline{w}_{j1}$ for $j = 1, 2$. Define

$$u_1 = H_1(q_{11}, q_{12}, p_1, p_2) = \frac{q_{11}p_1}{q_{11}p_1 + q_{12}p_2}, \tag{7.13}$$

and

$$u_2 = H_2(q_{11}, q_{12}, p_1, p_2) = \frac{q_{12}p_2}{q_{11}p_1 + q_{12}p_2}. \tag{7.14}$$

The quantities u_1 and u_2 are the quantities in equation (7.11) used to find the α-cuts of $\overline{w}_{11}$ and $\overline{w}_{21}$, respectively. If we find the min(max) of the expression in equation (7.11), then we have the end points of the interval $\overline{w}_{jk}[\alpha]$, see equation (2.27) and (2.28) in Chapter 2. We see that $\partial u_1/\partial q_{11} > 0$, $\partial u_1/\partial q_{12} < 0$, $\partial u_1/\partial p_1 > 0$ and $\partial u_1/\partial p_2 < 0$. Set $\overline{q}_{11}[\alpha] = [q_{111}(\alpha), q_{112}(\alpha)]$, $\overline{q}_{12}[\alpha] = [q_{121}(\alpha), q_{122}(\alpha)]$, $\overline{p}_1[\alpha] = [p_{11}(\alpha), p_{12}(\alpha)]$, $\overline{p}_2[\alpha] = [p_{21}(\alpha), p_{22}(\alpha)]$ and $\overline{w}_{11}[\alpha] = [w_{111}(\alpha), w_{112}(\alpha)]$. Then, because p_1, q_{11}, q_{12} are feasible

$$w_{111}(\alpha) = H_1(q_{111}(\alpha), q_{122}(\alpha), p_{11}(\alpha), p_{22}(\alpha)), \tag{7.15}$$

and

$$w_{112}(\alpha) = H_1(q_{112}(\alpha), q_{121}(\alpha), p_{12}(\alpha), p_{21}(\alpha)). \tag{7.16}$$

Similarly we compute $\overline{w}_{21}[\alpha]$ using u_2. Then

$$\overline{A}_{11}[\alpha] = 2\overline{w}_{11}[\alpha] + 5\overline{w}_{21}[\alpha], \tag{7.17}$$

for all α. Once we have $\overline{A}_{i1}$ for $i = 1, 2, 3$ we may obtain the best action. The graphs of these three fuzzy numbers are shown in Figures 7.1, 7.2 and 7.3. From these figures we conclude that, for $Z = z_1$, action a_3 is the best. In a similar manner we can do the computing when $Z = z_2$.

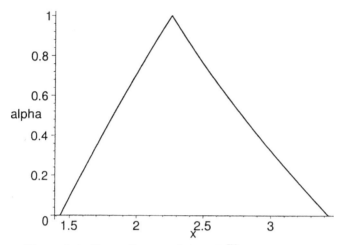

Figure 7.1: Fuzzy Expected Payoff $\overline{A}_{11}$, $Z = z_1$, in Example 7.3.1

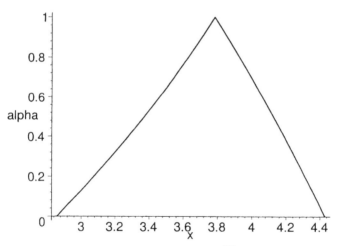

Figure 7.2: Fuzzy Expected Payoff $\overline{A}_{21}$, $Z = z_1$, in Example 7.3.1

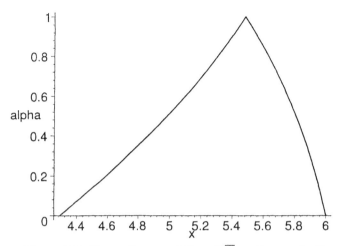

Figure 7.3: Fuzzy Expected Payoff $\overline{A}_{31}$, $Z = z_1$, in Example 7.3.1

7.4　References

1. J.G.Kemeny, A.Schleifer Jr.,J.L.Snell and G.L.Thompson: Finite Mathematics With Business Applications, Second Edition,Prentice-Hall, Englewoods, N.J., 1962.

2. B.W.Lindgren: Statistical Theory, Third Edition, Macmillan, N.Y., 1976.

3. R.D.Luce and H.Raiffa, Games and Decisions, John Wiley and Sons, N.Y., 1957.

4. H.A.Taha: Operations Research, Fifth Edition, Macmillan, N.Y., 1992.

Chapter 8

Continuous Fuzzy Random Variables

8.1 Introduction

This Chapter is based on [1]. New material includes the applications in Section 8.5. We consider the fuzzy uniform in Section 8.2, the fuzzy normal is in Section 8.3, followed by the fuzzy negative exponential in Section 8.4. In each case of a fuzzy density function we first discuss how they are used to compute fuzzy probabilities and then we find their fuzzy mean and their fuzzy variance. We always substitute fuzzy numbers for the parameters in these probability density functions, justified by Section 2.8, to produce fuzzy probability density functions.

We will denote the normal probability density as $N(\mu, \sigma^2)$ and the fuzzy normal density as $N(\overline{\mu}, \overline{\sigma}^2)$. The uniform density on interval $[a, b]$ is written $U(a, b)$ and the fuzzy uniform $U(\overline{a}, \overline{b})$ for fuzzy numbers $\overline{a}$ and $\overline{b}$. The negative exponential is $E(\lambda)$ with fuzzy form $E(\overline{\lambda})$.

8.2 Fuzzy Uniform

The uniform density $U(a, b)$, $a < b$, has $y = f(x; a, b) = 1/(b - a)$ for $a \leq x \leq b$ and $f(x; a, b) = 0$ otherwise. Now consider $U(\overline{a}, \overline{b})$ for fuzzy numbers $\overline{a}$ and $\overline{b}$. If $\overline{a}[1] = [a_1, a_2]$ and $\overline{b}[1] = [b_1, b_2]$ we assume that $a \in [a_1, a_2]$, $b \in [b_1, b_2]$ so that $\overline{a}$ ($\overline{b}$) represents the uncertainty in a (b). Now using the fuzzy uniform density we wish to compute the fuzzy probability of obtaining a value in the interval $[c, d]$. Denote this fuzzy probability as $\overline{P}[c, d]$. We can easily generalize to $\overline{P}[E]$ for more general subsets E.

There is uncertainty in the end points of the uniform density but there is no uncertainty in the fact that we have a uniform density. What this means is that given any $s \in \bar{a}[\alpha]$ and $t \in \bar{b}[\alpha]$, $s < t$, we have a $U(s,t)$, or $f(x; s, t) = 1/(t - s)$ on $[s, t]$ and it equals zero otherwise , for all $0 \leq \alpha \leq 1$. This enables us to find fuzzy probabilities. Let $L(c, d; s, t)$ be the length of the interval $[s, t] \cap [c, d]$. Then

$$\overline{P}[c, d][\alpha] = \{L(c, d; s, t)/(t - s)|s \in \bar{a}[\alpha], t \in \bar{b}[\alpha], s < t\}, \qquad (8.1)$$

for all $\alpha \in [0, 1]$. Equation (8.1) defines the α-cuts and we put these α-cuts together to obtain the fuzzy set $\overline{P}[c, d]$. To find an α-cut of $\overline{P}[c, d]$ we find the probability of getting a value in the interval $[c, d]$ for each uniform density $U(s, t)$ for all $s \in \bar{a}[\alpha]$ and all $t \in \bar{b}[\alpha]$, with $s < t$.

Example 8.2.1

Let $\bar{a} = (0/1/2)$ and $\bar{b} = (3/4/5)$ and $[c, d] = [1, 4]$. Now $\overline{P}[c, d][\alpha] = [p_1(\alpha), p_2(\alpha)]$ an interval whose end points are functions of α. Then $p_1(\alpha)$ is the minimum value of the expression on the right side of equation (8.1) and $p_2(\alpha)$ is the maximum value. That is

$$p_1(\alpha) = min\{L(1, 4; s, t)/(t - s)|s \in \bar{a}[\alpha], t \in \bar{b}[\alpha]\}, \qquad (8.2)$$

and

$$p_2(\alpha) = max\{L(1, 4; s, t)/(t - s)|s \in \bar{a}[\alpha], t \in \bar{b}[\alpha]\}. \qquad (8.3)$$

It is easily seen that $p_2(\alpha) = 1$ all α in this example. To find the minimum we must consider four cases. First $\bar{a}[\alpha] = [\alpha, 2 - \alpha]$ and $\bar{b}[\alpha] = [3 + \alpha, 5 - \alpha]$. Then the cases are: (1) $\alpha \leq s \leq 1, 3 + \alpha \leq t \leq 4$; (2) $\alpha \leq s \leq 1, 4 \leq t \leq 5 - \alpha$; (3) $1 \leq s \leq 2 - \alpha, 3 + \alpha \leq t \leq 4$; and (4) $1 \leq s \leq 2 - \alpha, 4 \leq t \leq 5 - \alpha$. Studying all four cases we obtain the minimum equal to $3/(5 - 2\alpha)$. Hence the α-cuts of $\overline{P}[1, 4]$ are $[3/(5 - 2\alpha), 1]$ and the graph of this fuzzy number is in Figure 8.1

Next we want to find the mean and variance of $U(\bar{a}, \bar{b})$. Let the mean be $\bar{\mu}$ and we find its α-cuts as follows

$$\bar{\mu}[\alpha] = \{\int_s^t (x/(t - s))dx|s \in \bar{a}[\alpha], t \in \bar{b}[\alpha], s < t\}, \qquad (8.4)$$

for all α. But each integral in equation (8.4) equals $(s+t)/2$. Hence, assuming $\bar{a}[0] = [s_1, s_2]$, $\bar{b}[0] = [t_1, t_2]$ and $s_2 < t_1$,

$$\bar{\mu} = (\bar{a} + \bar{b})/2. \qquad (8.5)$$

So, $\bar{\mu}$ is the fuzzification of the crisp mean $(a + b)/2$. If the variance of $U(\bar{a}, \bar{b})$ is $\bar{\sigma}^2$, then its α-cuts are

$$\bar{\sigma}^2[\alpha] = \{\int_s^t [(x - \mu)^2/(t - s)]dx|s \in \bar{a}[\alpha], t \in \bar{b}[\alpha], \mu = (s+t)/2, s < t\}, \qquad (8.6)$$

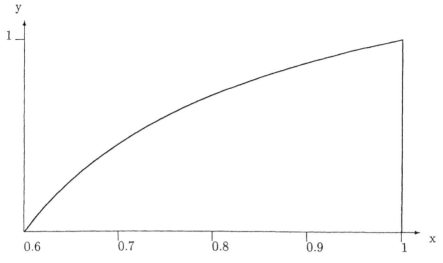

Figure 8.1: Fuzzy Probability in Example 8.2.1

for all α. Each integral in equation (8.6) equals $(t - s)^2/12$. Hence $\overline{\sigma}^2 = (\overline{b} - \overline{a})^2/12$, the fuzzification of the crisp variance.

Next we look at the fuzzy normal probability density.

8.3 Fuzzy Normal

The normal density $N(\mu, \sigma^2)$ has density function $f(x; \mu, \sigma^2)$, $x \in \mathbf{R}$, mean μ and variance σ^2. So consider the fuzzy normal $N(\overline{\mu}, \overline{\sigma}^2)$ for fuzzy numbers $\overline{\mu}$ and $\overline{\sigma}^2 > 0$. We wish to compute the fuzzy probability of obtaining a value in the interval $[c, d]$. We write this fuzzy probability as $\overline{P}[c, d]$. We may easily extend our results to $\overline{P}[E]$ for other subsets E of $\mathbf{R}$. For $\alpha \in [0, 1]$, $\mu \in \overline{\mu}[\alpha]$ and $\sigma^2 \in \overline{\sigma}^2[\alpha]$ let $z_1 = (c - \mu)/\sigma$ and $z_2 = (d - \mu)/\sigma$. Then

$$\overline{P}[c, d][\alpha] = \{\int_{z_1}^{z_2} f(x; 0, 1)dx | \mu \in \overline{\mu}[\alpha], \sigma^2 \in \overline{\sigma}^2[\alpha]\}, \qquad (8.7)$$

for $0 \leq \alpha \leq 1$. The above equation gets the α-cuts of $\overline{P}[c, d]$. Also, in the above equation $f(x; 0, 1)$ stands for the standard normal density with zero mean and unit variance. Let $\overline{P}[c, d][\alpha] = [p_1(\alpha), p_2(\alpha)]$. Then the minimum (maximum) of the expression on the right side of the above equation is $p_1(\alpha)$ ($p_2(\alpha)$). In general, it will be difficult to find these minimums (maximums) and one might consider using a genetic (evolutionary) algorithm, or some other numerical technique. However, as the following example shows, in some cases we can easily compute these α-cuts.

α	$\overline{P}[10,15][\alpha]$
0	[0.1584,0.7745]
0.2	[0.2168,0.7340]
0.4	[0.2821,0,6813]
0.6	[0.3512,0.6203]
0.8	[0.4207,0.5545]
1.0	[0.4873,0.4873]

Table 8.1: Alpha-cuts of the Fuzzy Probability in Example 8.3.1

Example 8.3.1

Suppose $\overline{\mu} = (8/10/12)$, or the mean is approximately 10, and $\overline{\sigma}^2 = (4/5/6)$, or the variance is approximately five. Compute $\overline{P}[10,15]$. First it is easy to find the $\alpha = 1$ cut and we obtain $\overline{P}[10,15][1] = 0.4873$. Now we want the $\alpha = 0$ cut. Using the software package Maple [3] we graphed the function

$$g(x,y) = \int_{z_1}^{z_2} f(u;0,1)du, \qquad (8.8)$$

for $z_1 = (10-x)/y$, $z_2 = (15-x)/y$, $8 \le x \le 12$, $4 \le y^2 \le 6$. Notice that the $\alpha = 0$ cut of $(8/10/12)$ is $[8,12]$, the range for $x = \mu$, and of $(4/5/6)$ is $[4,6]$ the range for $y^2 = \sigma^2$. The surface clearly shows: (1) a minimum of 0.1584 at $x = 8$ and $y = 2$; and (2) a maximum of 0.7745 at $x = 12$ and $y = 2$. Hence the $\alpha = 0$ cut of this fuzzy probability is $[0.1584, 0.7745]$. But from this graph we may also find other α-cuts. We see from the graph that $g(x,y)$ is an increasing function of: (1) x for y fixed at a value between 2 and $\sqrt{6}$; and (2) y for x fixed at 8. However, $g(x,y)$ is a decreasing function of y for $x = 12$. This means that for any α-cut: (1) we get the max at $y = $ its smallest value and $x = $ at its largest value; and (2) we have the min when $y = $ at is smallest and $x = $ its least value. Some α-cuts of $\overline{P}[10,15]$ are shown in Table 8.1 and Figure 8.2 displays this fuzzy probability. The graph in Figure 8.2 is only an approximation because we did not force the graph through all the points in Table 8.1.

We now show that the fuzzy mean of $N(\overline{\mu}, \overline{\sigma}^2)$ is $\overline{\mu}$ and the fuzzy variance is $\overline{\sigma}^2$, respectively, the fuzzification of the crisp mean and variance. Let the fuzzy mean be $\overline{M}$. Then its α-cuts are

$$\overline{M}[\alpha] = \{\int_{-\infty}^{\infty} xf(x;\mu,\sigma^2)dx | \mu \in \overline{\mu}[\alpha], \sigma^2 \in \overline{\sigma}^2[\alpha]\}. \qquad (8.9)$$

But the integral in the above equation equals μ for any $\mu \in \overline{\mu}[\alpha]$ and any $\sigma^2 \in \overline{\sigma}^2[\alpha]$. Hence $\overline{M} = \overline{\mu}$. Let the fuzzy variance be $\overline{V}$. Then its α-cuts are

$$\overline{V}[\alpha] = \{\int_{-\infty}^{\infty} (x-\mu)^2 f(x,\mu,\sigma^2)dx | \mu \in \overline{\mu}[\alpha], \sigma^2 \in \overline{\sigma}^2[\alpha], \}. \qquad (8.10)$$

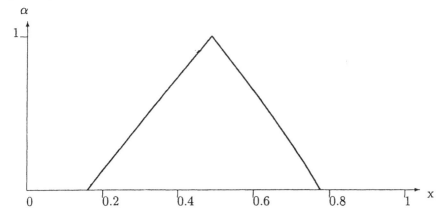

Figure 8.2: Fuzzy Probability in Example 8.3.1

We see that the integral in the above equation equals σ^2 for all $\mu \in \overline{\mu}[\alpha]$ and all $\sigma^2 \in \overline{\sigma}^2[\alpha]$. Therefore, $\overline{V} = \overline{\sigma}^2$.

8.4 Fuzzy Negative Exponential

The negative exponential $E(\lambda)$ has density $f(x; \lambda) = \lambda \exp(-\lambda x)$ for $x \geq 0$ and $f(x; \lambda) = 0$ otherwise, where $\lambda > 0$. The mean and variance of $E(\lambda)$ is $1/\lambda$ and $1/\lambda^2$, respectively. Now consider $E(\overline{\lambda})$ for fuzzy number $\overline{\lambda} > 0$. Let us find the fuzzy probability of obtaining a value in the interval $[c, d]$, $c > 0$. Denote this probability as $\overline{P}[c, d]$. One may generalize to $\overline{P}[E]$ for other subsets E of $\mathbf{R}$. We compute

$$\overline{P}[c, d][\alpha] = \{ \int_c^d \lambda \exp(-\lambda x) dx | \lambda \in \overline{\lambda}[\alpha] \}, \qquad (8.11)$$

for all α. Let $\overline{P}[c, d][\alpha] = [p_1(\alpha), p_2(\alpha)]$, then

$$p_1(\alpha) = min\{ \int_c^d \lambda \exp(-\lambda x) dx | \lambda \in \overline{\lambda}[\alpha] \}, \qquad (8.12)$$

and

$$p_2(\alpha) = max\{ \int_c^d \lambda \exp(-\lambda x) dx | \lambda \in \overline{\lambda}[\alpha] \}, \qquad (8.13)$$

for $0 \leq \alpha \leq 1$. Let

$$h(\lambda) = \exp(-c\lambda) - \exp(-d\lambda) = \int_c^d \lambda \exp(-\lambda x) dx, \qquad (8.14)$$

and we see that h: (1) is an increasing function of λ for $0 < \lambda < \lambda^*$; and (2) is a decreasing function of λ for $\lambda^* < \lambda$. We find that $\lambda^* = -[ln(c/d)]/(d -$

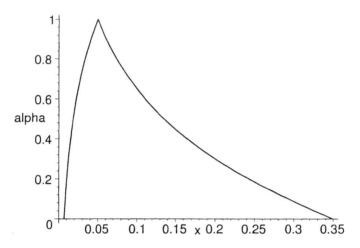

Figure 8.3: Fuzzy Probability for the Fuzzy Exponential

c). Assume that $\overline{\lambda} > \lambda^*$. So we can now easily find $\overline{P}[c, d]$. Let $\overline{\lambda}[\alpha] = [\lambda_1(\alpha), \lambda_2(\alpha)]$. Then

$$p_1(\alpha) = h(\lambda_2(\alpha)), \tag{8.15}$$

and

$$p_2(\alpha) = h(\lambda_1(\alpha)). \tag{8.16}$$

We give a picture of this fuzzy probability in Figure 8.3 when: (1) $c = 1$ and $d = 4$; and (2) $\overline{\lambda} = (1/3/5)$.

Next we find the fuzzy mean and fuzzy variance of $E(\overline{\lambda})$. If $\overline{\mu}$ denotes the mean, we find its α-cuts as

$$\overline{\mu}[\alpha] = \{\int_0^\infty x\lambda \exp(-\lambda x)dx | \lambda \in \overline{\lambda}[\alpha]\}, \tag{8.17}$$

for all α. However, each integral in the above equation equals $1/\lambda$. Hence $\overline{\mu} = 1/\overline{\lambda}$. If $\overline{\sigma}^2$ is the fuzzy variance, then we write down an equation to find its α-cuts we obtain $\overline{\sigma}^2 = 1/\overline{\lambda}^2$. The fuzzy mean (variance) is the fuzzification of the crisp mean (variance).

8.5 Applications

In this section we look at some applications of the fuzzy uniform, the fuzzy normal and the fuzzy negative exponential.

8.5.1 Fuzzy Uniform

Customers arrive randomly at a certain shop. Given that one customer arrived during a particular T-minute period, let X be the time within the T minutes that the customer arrived. Assume that the probability density function for X is $U(0, T)$. Find $Prob(4 \leq X \leq 9)$. However, T is not known exactly and is approximately 10, so we will use $\overline{T} = (8/10/12)$ for T. So the probability that $4 \leq X \leq 9$ becomes a fuzzy probability $\overline{P}[4, 9]$. Its α-cuts are computed as in equation (8.1). We find that for $0 \leq \alpha \leq 0.5$ that

$$\overline{P}[4, 9][\alpha] = \{\frac{min\{t, 9\} - 4}{t} | t \in [8 + 2\alpha, 12 - 2\alpha]\}, \tag{8.18}$$

and for $0.5 \leq \alpha \leq 1$,

$$\overline{P}[4, 9][\alpha] = \{\frac{5}{t} | t \in [8 + 2\alpha, 12 - 2\alpha]\}. \tag{8.19}$$

From this we determine that

$$\overline{P}[4, 9][\alpha] = [\frac{5}{12 - 2\alpha}, \frac{5}{9}], \tag{8.20}$$

for $0 \leq \alpha \leq 0.5$, and

$$\overline{P}[4, 9][\alpha] = [\frac{5}{12 - 2\alpha}, \frac{5}{8 + 2\alpha}], \tag{8.21}$$

for $0.5 \leq \alpha \leq 1$. The graph of this fuzzy probability is in Figure 8.4.

8.5.2 Fuzzy Normal Approximation to Fuzzy Binomial

We first review some basic information about the fuzzy binomial distribution from Chapter 4. Define $X = \{x_1, ..., x_n\}$ and let E be a non-empty, proper, subset of X. We have an experiment where the result is considered a "success" if the outcome x_i is in E. Otherwise, the result is considered a "failure". Let $P(E) = p$ so that $P(E') = q = 1 - p$. $P(E)$ is the probability of success and $P(E')$ is the probability of failure. We assume that $0 < p < 1$.

Suppose we have m independent repetitions of this experiment. If $P(r)$ is the probability of r successes in the m experiments, then

$$P(r) = \binom{m}{r} p^r (1 - p)^{m-r}, \tag{8.22}$$

for $r = 0, 1, 2, ..., m$, gives the binomial distribution. We write $b(m; p)$ for the crisp binomial and $b(m; \overline{p})$ for the fuzzy binomial. Throughout this section we are using $q = 1 - p$ which is different from the discussion of the fuzzy binomial in Chapter 4.

In these experiments let us assume that $P(E)$ is not known precisely and it needs to be estimated, or obtained from expert opinion. So the p value is

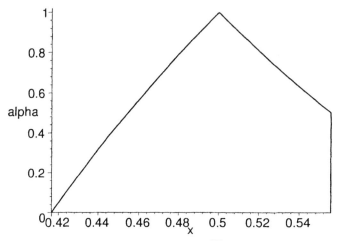

Figure 8.4: Fuzzy Probability $\overline{P}[4,9]$ for the Fuzzy Uniform

uncertain and we substitute $\bar{p}$ for p. Now let $\overline{P}(r)$ be the fuzzy probability of r successes in m independent trials of the experiment. Then

$$\overline{P}(r)[\alpha] = \{\binom{m}{r}p^r(1-p)^{m-r}|p \in \bar{p}[\alpha]\}, \qquad (8.23)$$

for $0 \le \alpha \le 1$. If $\overline{P}(r)[\alpha] = [P_{r1}(\alpha), P_{r2}(\alpha)]$, then

$$P_{r1}(\alpha) = min\{\binom{m}{r}p^r(1-p)^{m-r}|p \in \bar{p}[\alpha]\}, \qquad (8.24)$$

and

$$P_{r2}(\alpha) = max\{\binom{m}{r}p^r(1-p)^{m-r}|p \in \bar{p}[\alpha]\}. \qquad (8.25)$$

Example 8.5.2.1

Let $p = 0.4$ and $m = 3$. Since p is uncertain we use $\bar{p} = (0.3/0.4/0.5)$ for p. Now we will calculate the fuzzy number $\overline{P}(2)$. Equations (8.24) and (8.25) become

$$P_{r1}(\alpha) = min\{3p^2(1-p)|p \in \bar{p}[\alpha]\}, \qquad (8.26)$$

and

$$P_{r2}(\alpha) = max\{3p^2(1-p)|p \in \bar{p}[\alpha]\}. \qquad (8.27)$$

Since $d(3p^2(1-p))/dp > 0$ on $\bar{p}[0]$ we obtain

$$\overline{P}(2)[\alpha] = [3(p_1(\alpha))^2(1-p_1(\alpha)), 3(p_2(\alpha))^2(1-p_2(\alpha))], \qquad (8.28)$$

where $\bar{p}[\alpha] = [p_1(\alpha), p_2(\alpha)] = [0.3 + 0.1\alpha, 0.5 - 0.1\alpha]$.

We now need the mean and variance of the fuzzy binomial distribution $b(m; \bar{p})$ which was discussed in Section 4.2. Let $\bar{\mu}$ be the fuzzy mean of the fuzzy binomial and let $\bar{\sigma}^2$ be its fuzzy variance. We showed that, in general, $\bar{\mu} \leq m\bar{p}$ and $\bar{\sigma}^2 \leq m\bar{p}(1 - \bar{p})$. But when we use $q = 1 - p$ we obtain $\bar{\mu} = m\bar{p}$.

Now consider $b(100; \bar{p})$ and we wish to find the fuzzy probability of obtaining from 40 to 60 successes. Denote this fuzzy probability as $\overline{P}[40, 60]$ and direct calculation would be

$$\overline{P}[40, 60][\alpha] = \{\sum_{i=40}^{60} \binom{100}{i} p^i (1 - p)^{100-i} | p \in \bar{p}[\alpha]\}, \qquad (8.29)$$

for each α-cut. Now let us try to use the fuzzy normal to approximate this fuzzy probability. Let $f(x; 0, 1)$ be the normal probability density function with zero mean and unit variance. In the following equation $z_1 = (39.5-\mu)/\sigma$, $z_2 = (60.5 - \mu)/\sigma$, , then

$$\overline{P}[40, 60][\alpha] \approx \{\int_{z_1}^{z_2} f(x; 0, 1)dx | \mu \in \bar{\mu}[\alpha], \sigma^2 \in \bar{\sigma}^2[\alpha]\}, \qquad (8.30)$$

for all α, where $\bar{\mu}$ is the fuzzy mean of the fuzzy binomial and $\bar{\sigma}^2$ is the fuzzy variance of the fuzzy binomial. Let us show that equation (8.30) is correct through the following example.

Example 8.5.2.2

Let $m = 100$, $p \approx 0.6$ so that we use $\bar{p} = (0.5/0.6/0.7)$. For the normal approximation to the binomial to be reasonably accurate one usually assumes that [2] $mp > 5$ and $m(1 - p) > 5$. For the fuzzy normal approximation to the fuzzy binomial to be reasonably good we assume that $m\bar{p} > 5$ and $m(1 - \bar{p}) > 5$, which is true in this example. We now argue that equation (8.30) will give a good approximation to $\overline{P}[40, 60]$. Pick and fix a value of α in $[0, 1)$. Choose $p_0 \in \bar{p}[\alpha]$. Let

$$w = \sum_{i=40}^{60} \binom{100}{i} p_0^i (1 - p_0)^{100-i}, \qquad (8.31)$$

with $w \in \overline{P}[40, 60][\alpha]$.

Now we need to compute the fuzzy mean and the fuzzy variance of this fuzzy binomial. We get $\bar{\mu} = 100\bar{p}$. We next compute $\bar{\sigma}^2$ as in Example 4.2.2. We obtain

$$\bar{\sigma}^2[\alpha] = [h(p_2(\alpha), h(p_1(\alpha))], \qquad (8.32)$$

α	$\overline{P}[40,60][\alpha]$	Normal Approximation
0	[0.0210,0.9648]	[0.0191,0.9780]
0.2	[0.0558,0.9500]	[0.0539,0.9621]
0.4	[0.1235,0.9025]	[0.1228,0.9139]
0.6	[0.2316,0.8170]	[0.2329,0.8254]
0.8	[0.3759,0.6921]	[0.3786,0.6967]
1.0	[0.5379,0.5379]	[0.5406,0.5406]

Table 8.2: Fuzzy Normal Approximation to Fuzzy Binomial

where $h(p) = 100p(1-p)$, and $\overline{p}[\alpha] = [p_1(\alpha), p_2(\alpha)] = [0.5 + 0.1\alpha.0.7 - 0.1\alpha]$. The result is $\overline{\sigma}^2[\alpha] = [21 + 4\alpha - \alpha^2, 25 - \alpha^2]$. Then the α-cuts for $\overline{\sigma}$ will be the square root of the α-cuts of $\overline{\sigma}^2$.

Now let $\mu_0 = 100p_0$ in $100\overline{p}[\alpha]$ and let $\sigma_0 \in \overline{\sigma}[\alpha]$ which was computed above. Then

$$w \approx \int_{z_1}^{z_2} f(x; 0, 1)dx, \tag{8.33}$$

where $z_1 = (39.5 - \mu_0)/\sigma_0$, $z_2 = (60.5 - \mu_0)/\sigma_0$.

Now we turn it around and first pick $\mu_0 \in 100\overline{p}[\alpha]$ and $\sigma_0 \in \overline{\sigma}[\alpha]$. But this determines a $p_0 \in \overline{p}[\alpha]$, which then gives a value for w in equation(8.31). The approximation in equation (8.30) now holds.

So we see that under reasonable assumptions the fuzzy normal can approximate the fuzzy binomial. Table 8.2 shows the approximation for $\alpha = 0, 0.2, 0.4, 0.6, 0.8, 1$. Let us explain how we determined the values in Table 8.2. First we graphed the function

$$H(p) = \sum_{x=40}^{60} \binom{100}{x} p^x (1-p)^{100-x}, \tag{8.34}$$

for $p \in [0.5, 0.7]$ and found it is a decreasing function of p on this interval. We then easily found the α-cuts for the fuzzy binomial in Table 8.2. We calculated the α-cuts for the fuzzy normal using the "graphical" method described in Section 2.9.

8.5.3   Fuzzy Normal Approximation to Fuzzy Poisson

The fuzzy Poisson was discussed in Section 4.3. Let X be a random variable having a Poisson probability mass function so that, if $P(x)$ is the probability that $X = x$, we have $P(x) = \lambda^x \exp(-\lambda)/x!$, for $x = 0, 1, 2, 3...$ and $\lambda > 0$. We know, if λ is sufficiently large [2], that we can approximate the crisp Poisson with the crisp normal. Let $\lambda = 20$ and let $P(16, 21]$ be the probability that $16 < X \leq 21$. Then

$$P(16, 21] \approx \int_{z_1}^{z_2} f(x; 0, 1)dx, \tag{8.35}$$

α	$\overline{P}(16,21][\alpha]$	Fuzzy Normal Approximation
0	[0.1868,0.4335]	[0.1690,0.4814]
0.2	[0.2577,0.4335]	[0.2356,0.4814]
0.4	0.3073,0.4335]	[0.2896,0.4814]
0.6	[0.3546,0.4335]	[0.3371,0.4814]
0.8	[0.3948,0.4335]	[0.3804,0.4814]
1	[0.4226,0.4226]	[0.4144,0.4144]

Table 8.3: Fuzzy Normal Approximation to Fuzzy Poisson

where $z_1 = (16.5 - \lambda)/\sqrt{\lambda}, z_2 = (21.5 - \lambda)/\sqrt{\lambda}$, and $f(x; 0, 1)$ is the normal probability density function with mean zero and variance one. We used that fact that the mean and variance of the crisp Poisson are both equal to λ to define the z_i. In equation (8.35) the exact value using the Poisson is 0.4226 and the normal approximation gives 0.4144. We now argue that we may use the fuzzy normal to approximate the fuzzy Poisson.

Example 8.5.3.1

Let $\overline{\lambda} = (15/20/25)$ and denote the fuzzy probability that $16 < X \leq 21$ as $\overline{P}(16, 21]$ whose α-cuts are

$$\overline{P}(16, 21][\alpha] = \{ \sum_{x=17}^{21} \lambda^x \exp(-\lambda)/x! | \lambda \in \overline{\lambda}[\alpha] \}, \tag{8.36}$$

for all α in $[0, 1]$. In the following equation $z_1 = (16.5 - \lambda)/\sqrt{\lambda}$ and $z_2 = (21.5 - \lambda)/\sqrt{\lambda}$, then

$$\overline{P}(16, 21][\alpha] \approx \{ \int_{z_1}^{z_2} f(x; 0, 1)dx | \lambda \in \overline{\lambda}[\alpha] \}, \tag{8.37}$$

for all α. The argument that this equation is correct is the same as that used in the previous subsection for the fuzzy binomial and the fuzzy normal. Table 8.3 shows the approximation for $\alpha = 0, 0.2, 0.4, 0.6, 0.8, 1$. We used the "graphical" method, described in Section 2.9, to estimate the α-cuts in Table 8.3. We notice that in this example the approximation is not too good. Perhaps, we need to consider a larger fuzzy value for $\overline{\lambda}$.

8.5.4   Fuzzy Normal

This example has been adapted from an example in [4]. Cockpits in fighter jets were originally designed only for men. However, the US Air Force now recognizes that women also make perfectly good pilots of fighter jets. So various cockpit changes were required to better accommodate the new women

α	$\overline{P}[140, 200][\alpha]$
0	[0.5000,0.9412]
0.2	[0.5538,0.9169]
0.4	[0.6083,0,8869]
0.6	[0.6622,0.8511]
0.8	[0.7146,0.8100]
1.0	[0.7642,0.7642]

Table 8.4: Alpha-cuts of the $\overline{P}[140, 200]$

pilots. The ejection seat used in the fighter jets was originally designed for men who weighted between 140 and 200 pounds. Based on the data they could get on the pool of possible new women pilots their weight was approximated normally distributed with estimated mean of 143 pounds having an estimated standard deviation of 25 pounds. Any women weighing less than 140 pounds, or more than 200 pounds, would have a greater chance of injury if they had to eject. So the US Air Force wanted to know , given a random sample on n possible women pilots, what is the probability that their mean weight is between 140 and 200 pounds. Answers to such questions are important for the possible redesign of the ejection seats.

The mean of 140 pounds, with standard deviation of 25 pounds, are point estimates and to use just these numbers will not show the uncertainty in these estimates. So we will instead use a set of confidence intervals, as described in Section 2.8, to construct fuzzy numbers $\overline{\mu}$, for the mean, and $\overline{\sigma}$, for the standard deviation. Assume $\overline{\mu} = (140/143/146)$ and $\overline{\sigma} = (23/25/27)$. Suppose y is the mean of the weights of the random sample of $n = 36$ possible women pilots. We now want to calculate the fuzzy probability $\overline{P}[140, 200]$ that $140 \leq y \leq 200$ for y having the fuzzy normal with mean $\overline{\mu}$ and standard deviation $\overline{\sigma}/\sqrt{36}$. We therefore need to calculate the α-cuts

$$\overline{P}[140, 200][\alpha] = \{ \int_{z_1}^{z_2} f(x; 0, 1)dx | \mu \in \overline{\mu}[\alpha], \sigma \in \overline{\sigma}[\alpha]\}, \qquad (8.38)$$

all α, where $z_1 = 6(140 - \mu)/\sigma$ and $z_2 = 6(200 - \mu)/\sigma$. The value of equation (8.38) is easily found for $\alpha = 1$ and it is 0.7642 . Also, as in Example 8.3.1 we can get the value when $\alpha = 0$. We used the "graphical" method discussed in Section 2.9 to estimate the α-cuts in Table 8.4. The graph of this fuzzy probability is shown in Figure 8.5. The graph in Figure 8.5 is not completely accurate because we did not force it to go through all the points given in Table 8.4.

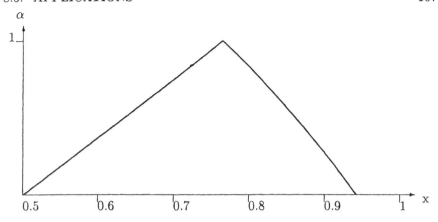

Figure 8.5: Fuzzy Probability in the Ejection Seat Example

8.5.5 Fuzzy Negative Exponential

The crisp negative exponential probability density function is related to the crisp Poisson probability mass function and the same is true in the fuzzy case. A machine has a standby unit available for immediate replacement upon failure. Assume that failures occur for these machines at a rate of λ per hour. Let X be a random variable which counts the number of failures during a period of T hours. Assume that X has a Poisson probability mass function and the probability that $X = x$, denoted by $P_T(x)$, is

$$P_T(x) = (\lambda T)^x \exp(-\lambda T)/x!, \tag{8.39}$$

for $x = 0, 1, 2, 3...$ Now let Y be the random variable whose value is the waiting time to the first failure. It is well known [2] that Y has the exponential probability density function so that

$$Prob[Y > t] = \int_t^\infty \lambda \exp(-\lambda x) dx, \tag{8.40}$$

which is the probability that the first failure occurs after t hours.

Now switch to the fuzzy Poisson with $\overline{\lambda} = (0.07/0.1/0.13)$ for λ and denote the fuzzy probability that the first failure occurs after 10 hours as $\overline{P}[10, \infty]$. Then its α-cuts are

$$\overline{P}[10, \infty][\alpha] = \{ \int_{10}^\infty \lambda \exp(-\lambda x) dx | \lambda \in \overline{\lambda}[\alpha] \}, \tag{8.41}$$

for all α. These α-cuts are easy to find because the integral in the above equation is simply $\exp(-10\lambda)$ for $\lambda \in \overline{\lambda}[\alpha]$. So, if $\overline{\lambda}[\alpha] = [\lambda_1(\alpha), \lambda_2(\alpha)]$, then

$$\overline{P}[10, \infty][\alpha] = [\exp(-10\lambda_2(\alpha)), \exp(-10\lambda_1(\alpha))], \tag{8.42}$$

which equals

$$\overline{P}[10, \infty][\alpha] = [\exp(-1.3 + 0.3\alpha), \exp(-0.7 - 0.3\alpha)]. \qquad (8.43)$$

To summarize, if we substitute $\overline{\lambda}$ for λ in Equation (8.39) the fuzzy Poisson can be used to find the fuzzy probability of x failures in time interval T and the fuzzy negative exponential gives the fuzzy times between successive failures. An important property of the crisp exponential is its "forgetfullness". The probability statement of this property is

$$Prob[Y > t_1 + t_2 | Y > t_2] = Prob[Y > t_1]. \qquad (8.44)$$

The time interval remaining until the next failure is independent of the time interval that has elapsed since the last failure. We now show this is also true for the fuzzy exponential using our definition of fuzzy conditional probability in Section 3.3. Using the fuzzy negative exponential α-cuts of the fuzzy conditional probability, relating to the left side of equation (8.44), is

$$\overline{P}[Y > t_1 + t_2 | Y > t_2][\alpha] = \{ \frac{\int_{t_1+t_2}^{\infty} \lambda \exp(-\lambda x) dx}{\int_{t_2}^{\infty} \lambda \exp(-\lambda x) dx} | \lambda \in \overline{\lambda}[\alpha] \}, \qquad (8.45)$$

for $\alpha \in [0, 1]$. Now the quotient of the integrals in equation (8.45) equals, after evaluation, $\exp(-t_1\lambda)$, so

$$\overline{P}[Y > t_1 + t_2 | Y > t_2][\alpha] = \{ \int_{t_1}^{\infty} \lambda \exp(-\lambda x) dx | \lambda \in \overline{\lambda}[\alpha] \}, \qquad (8.46)$$

which equals $\overline{P}[Y > t_1][\alpha]$. Hence, equation (8.44) also holds for the fuzzy negative exponential and it has the "forgetfullness" property.

8.6   References

1. J.J.Buckley and E.Eslami: Uncertain Probabilities II: The Continuous Case. Under review.

2. R.V.Hoog and E.A.Tanis: Probability and Statistical Inference, Sixth Edition, Prentice Hall, Upper Saddle River, N.J., 2001.

3. Maple 6, Waterloo Maple Inc. Waterloo, Canada.

4. M.F.Triola: Elementary Statistics with Excel, Addison-Wesley, N.Y., 2001.

Chapter 9

Fuzzy Inventory Control

9.1   Introduction

Most of the applications of fuzzy sets to inventory control have been in the area of the EOQ, or the economic order quantity problem. See ([4]-[12],[14]-[16]), and the references in these papers, for a survey of this literature. What was usually done by these authors is to fuzzify some of the, difficult to estimate, parameters in the model. We are going to do something different in this Chapter in modeling demand with a fuzzy normal probability density. Our previous research in this area includes using fuzzy sets to model uncertainty in a single item, N-period, inventory control problem ([2],[3]). In that research we had demand (not a fuzzy normal) crisp or fuzzy, ordering cost and holding cost fuzzy, with or without shortages and backorders.

We will first look at, in the next section, a simple single item inventory model with probabilistic demand using the decision criterion of minimizing expected costs. This model is based on a fuzzy inventory control problem studied in [1] and may be found in most operations research/managemant science books, and we have adapted ours from ([13], p.519-522). We first discuss the non-fuzzy model and then the fuzzy model. Then in Section 9.3 we extend this model to multiple periods.

9.2   Single Period Model

Our model is a single period model where an item is ordered only once to satisfy demand for a specific period. Let x_0 be the amount on hand before an order is placed. Stock replacement occurs instantaneously after an order is placed. The amount on hand after the order is received is y so $y - x_0$ is the size of the order. If no items are ordered , then $y = x_0$. We also assume instantaneous demand, total demand is filled at the start of the period. D stands for the demand and $f(D; \mu, \sigma^2)$ is the probability density of demand

which is the normal density with mean μ and variance σ^2. So at the start
of the period we have x_0 items, we order more items or we do not order and
we now have y units on hand at the start of the period and then we receive
a demand of D units leaving us with a surplus of $y - D$ or a shortage of
$D - y$. Let h be the storage cost per unit per period, p is the shortage cost
per unit per period and c is the purchase cost per unit. Assume $p > c$ (see
equation (9.2)). We assume there is no order (set-up) costs. $H(y)$ is the
holding inventory which equals $y - D$ if $D < y$ and is zero otherwise. $S(y)$
will be the shortage inventory which equals $D - y$ when $D \geq y$ and is zero
otherwise. If $E(y)$ is the expected cost per period, a function of y, then it is
calculated as follows

$$E(y) = c(y - x_0) + h \int_0^y (y - D)f(D; \mu, \sigma^2)dD + p \int_y^\infty (D - y)f(D; \mu, \sigma^2)dD.$$
$$(9.1)$$

The optimal y value, written y^*, is defined by the equation

$$\int_0^{y^*} f(D; \mu, \sigma^2)dD = \frac{p - c}{p + h}.$$
$$(9.2)$$

The optimal ordering policy is given by: (1) order $y^* - x_0$ if $y^* > x_0$; and (2)
do not order when $y^* \leq x_0$.

It is common to model demand with a normal density but μ and σ^2 must
be estimated from the data. Hence, the two parameters are uncertain and
we will substitute fuzzy numbers for them giving the fuzzy normal $N(\overline{\mu}, \overline{\sigma}^2)$,
see Section 2.8. $E(y)$ becomes a fuzzy number whose α-cuts are

$$\overline{E}(y)[\alpha] = \{E(y)|\mu \in \overline{\mu}[\alpha], \sigma^2 \in \overline{\sigma}^2[\alpha]\},$$
$$(9.3)$$

for all α. Now we get a fuzzy number $\overline{E}(y)$ for all $y \geq x_0$ and we want to
find the optimal y value y^* to minimize $\overline{E}(y)$.

We can not minimize a fuzzy number so what we are going to do, which
we have done before (see Section 2.5), is first change $min\overline{E}(y)$ into a multiobjective problem and then translate the multiobjective problem into a single
objective problem. This strategy is adopted from the finance literature where
they had the problem of minimizing a random variable X whose values are
constrained by a probability density function. They considered the multiobjective problem: (1) minimize the expected value of X; (2) minimize the
variance of X; and (3) minimize the skewness of X to the right of the expected value. For our problem let: (1) $c(y)$ be the center of the core of $\overline{E}(y)$,
the core of a fuzzy number is the interval where the membership function
equals one, for each $y \geq x_0$; (2) $L(y)$ be the area under the graph of the
membership function to the left of $c(y)$; and (3) $R(y)$ be the area under the
graph of the membership function to the right of $c(y)$. See Figure 2.5. For
$min\overline{E}(y)$ we substitute: (1) $min[c(y)]$; (2) $maxL(y)$, or maximize the possibility of obtaining values less than $c(y)$; and (3) $minR(y)$, or minimize the

possibility of obtaining values greater then $c(y)$. The single objective problem is then

$$min(\lambda_1[M - L(y)] + \lambda_2 c(y) + \lambda_3 R(y)), \qquad (9.4)$$

where $\lambda_i > 0$, $1 \leq i \leq 3$, $\lambda_1 + \lambda_2 + \lambda_3 = 1$, and $M > 0$ is sufficiently large so that $maxL(y)$ is equivalent to $min[M - L(y)]$. The decision maker is to choose the values of the weights λ_i for the three goals. Usually one picks different values for the λ_i to explore the solution set and then lets the decision maker choose y^* from this set of solutions.

We can easily handle fuzzy values for h and p. Holding cost and lost sales cost are always difficult to estimate and could be modeled as fuzzy numbers. The only change would be to substitute $\overline{h}[\alpha]$ for h and substitute $\overline{p}[\alpha]$ for p in equation (9.1), which then goes into equation (9.3). In equation (9.3) we would then add $h \in \overline{h}[\alpha]$ and $p \in \overline{p}[\alpha]$.

9.3 Multiple Periods

We will employ the same notation and assumptions as in the previous section. However, now we: (1) have N periods with the assumption that final inventory is zero ($x_{N+1} = 0$); (2) will sell the item at \$r/unit so we want to maximize profit; and (3)will discount future monies back to their present value.

In the ith period incoming inventory is x_i and after receiving our order we have on hand y_i units and demand is D. If $y_i > x_i$, then the order was $y_i - x_i$, and if $y_i = x_i$, then there was no order. The starting inventory is now x_1 which is known. The out going inventory x_{i+1}, which is the starting inventory for the next period, is $y_i - D$ when $y_i > D$ and is zero when $D \geq y_i$.

The expected profit for the ith period is

$$
\begin{aligned}
E(y_i) &= -c(y_i - x_i) + \int_0^{y_i} [rD - h(y_i - D)]f(D; \mu, \sigma^2)dD \\
&+ \int_{y_i}^{\infty} [ry_i - p(D - y_i)]f(D; \mu, \sigma^2)dD,
\end{aligned}
\qquad (9.5)
$$

where $f(D; \mu, \sigma^2)$ is the normal density. The total expected profit is then

$$TP(y) = \sum_{i=1}^{N} (\tau)^{i-1} E(y_i), \qquad (9.6)$$

where τ discounts future money back to its present value and $y = (y_1, ..., y_N)$ is the vector of decision variables. So $\tau = (1 + \beta)^{-1}$ for some interest rate β. The object is to determine y to maximize $TP(y)$. Usually dynamic programming is used to solve the problem.

We will need a method of calculating the x_i values since $x_{i+1} = y_i - D$ when $y_i > D$ and equals zero otherwise, and D is a random variable in the crisp case or a fuzzy random variable in the fuzzy case. In the crisp case we could set $x_{i+1} = max\{0, E[y_i - D]\}$, $1 \le i \le N$, where "E" denotes expected value. Since $E[y_i] = y_i$ and $E[D] = \mu$, the mean of demand, we have in the crisp case $x_{i+1} = max\{0, y_i - \mu\}$. When demand is given by a fuzzy normal we would use the expression $x_{i+1} = max\{0, defuzzifiedE[y_i - D]\}$ because now $E[y_i - D] = y_i - \overline{\mu}$ which is a fuzzy set. Our policy in this model is that everything is crisp except demand. Hence we must assign an integer to the fuzzy set $y_i - \overline{\mu}$ which is accomplished by the command "defuzzify". Defuzzify is a method of assigning a real number to a fuzzy set [2], and then round that real number to an integer value.

As in Section 9.2 the fuzzy problem arrives when we use the fuzzy normal in equation (9.5). Alpha-cuts of the fuzzy expected profit $\overline{E}(y_i)$ are computed as in equation (9.3). The fuzzy total profit $\overline{TP}(y)$ is calculated as in equation (9.6) using the $\overline{E}(y_i)$. Then we set up a single objective problem as in equation (9.4) to solve, but with a constraint $x_{N+1} = 0$. There will be slight changes in equation (9.4) because we now wish to find y to maximize the objective function. This has become a rather complicated problem and we recommend using an evolutionary algorithm, and not fuzzy dynamic programming, to obtain good approximate solutions [3]. We may also generalize to fuzzy numbers for the x_i, $2 \le i \le N + 1$, and fuzzy numbers for h and p as discussed in Section 9.2. If x_{N+1} is fuzzy , then we must change the constraint $x_{N+1} = 0$ to $\overline{x}_{N+1} \approx 0$, see [3].

The solution will give optimal values for the y_i, say y_i^*. Then the optimal ordering policy would be: (1) if $x_i < y_i^*$, the order $y_i^* - x_i$; and (2) if $x_i \ge y_i^*$, do not order. However, if N is not small, the the evolutionary algorithm will take a long time to run because most of the time is spent checking the constraint. What is needed is a feasible set of y_i values and the corresponding x_i values with $x_{N+1} = 0$. We always start with a given value of x_1. Then choose a value of y_1 and compute x_2. Next choose y_2, compute x_3, etc. and the set of y_i are feasible when we get $x_{N+1} = 0$. This checking for feasible y_i goes on with choosing the initial population and during the crossover and mutation operations. Once we have a feasible set of y_i values we may compute the α-cuts of $\overline{E}(y_i)$ and then α-cuts of $\overline{TP}(y)$. Then we have the (approximate) shape of the fuzzy number $\overline{TP}(y)$ and we can go on to determine $c(y)$, $L(y)$ and $R(y)$ to get the optimization problem

$$max(\lambda_1[M - L(y)] + \lambda_2 c(y) + \lambda_3 R(y)), \qquad (9.7)$$

for given values of the λ_i. This is a difficult problem to solve using fuzzy dynamic programming or an evolutionary algorithm. It is also difficult to solve in the crisp case using dynamic programming [13].

9.4   References

1. J.J.Buckley and E.Eslami: Uncertain Probabilities II: The Continuous Case. Under Review.

2. J.J.Buckley, E.Eslami and T.Feuring: Fuzzy Mathematics in Economics and Engineering, Physica-Verlag, Heidelberg, Germany, 2002.

3. J.J.Buckley, T.Feuring and Y.Hayashi: Solving Fuzzy Problems in Operations Research: Inventory Control, Soft Computing. To appear.

4. S.-C.Chang: Fuzzy Production Inventory for Fuzzy Product Quantity with Triangular Fuzzy Numbers, Fuzzy Sets and Systems, 107(1999), pp. 37-57.

5. M.Gen, Y.Tsujiomura and D.Zheng: An Application of Fuzzy Set Theory to Inventory Control Problems, Computers in Industrial Engineering, 33 (1997), pp. 553-556.

6. A.L.Guiffrida and R.Nagi: Fuzzy Set Theory Applications in Production Management Research: A Literature Survey, J. Intelligent Manufacturing, 9(1998), pp. 39-56.

7. J.Kacprzyk: Multistage Decision-Making under Fuzziness, Verlag TÜV Reinland, Köln, Germany, 1983.

8. H.Katagiri and H.Ishii: Some Inventory Problems with Fuzzy Shortage Cost, Fuzzy Sets and Systems, 111(2000), pp. 87-97.

9. H.-M.Lee and J.-S.Yao: Economic Order Quantity in Fuzzy Sense for Inventory without Backorder Model, Fuzzy Sets and Systems, 105(1999), pp. 13-31.

10. D.-C.Lin and J.-S.Tao: Fuzzy Economic Production for Production Inventory, Fuzzy Sets and Systems, 111(2000), pp. 465-495.

11. B.L.Liu and A.O.Esogbue: Decision Criteria and Optimal Inventory Processes, Kluwer Academic Publishers, Norwell, Mass., 1999.

12. G.Sommer: Fuzzy Inventory Scheduling in Applied Systems, in; G.E.Lasker (ed.), Applied Systems and Cybernetics, Vol. VI, Pergamon Press, N.Y., 1981, pp. 3052-3062.

13. H.A.Taha: Operations Research, Fifth Edition, Macmillan, N.Y., 1992.

14. J.-S.Yao and H.-M.Lee: Fuzzy Inventory with or without Backorder Quantity with Trapezoidal Fuzzy Numbers, Fuzzy Sets and Systems, 105(1999), pp. 311-337.

15. J.-S.Yao and J.-S.Su: Fuzzy Inventory with Backorder for Fuzzy Total Demand Based on Interval–Valued Fuzzy Set, European J. Operational Research, 124 (2000), pp. 390-408.

16. J.-S.Yao, S.-C.Shang and J.-S.Su: Fuzzy Inventory without Backorder for Fuzzy Order Quantity and Fuzzy Total Demand Quantity, Computers and Operations Research, 27 (2000), pp. 935-962.

Chapter 10

Joint Fuzzy Probability Distributions

10.1 Introduction

This Chapter generalizes Chapters 4 and 8 to multivariable fuzzy probability distributions. Since the discrete case and the continuous case are similar, just interchange summation and integrals symbols, we only present the continuous case in Section 10.2. Applications are in the next Chapter. The continuous case is based on [1]. For simplicity we will consider only the joint fuzzy probability distributions for two fuzzy random variables.

10.2 Continuous Case

Let X and Y be two random variables having joint probability density $f(x, y; \theta)$, where $x \in \mathbf{R}$ and $\theta = (\theta_1, ... \theta_n)$ is the vector of parameters defining the joint density. Usually we estimate these parameters using a random sample from the population. These estimates can be a point estimate or a confidence interval. We would like to substitute a confidence interval for each θ_i, instead of a point estimate, into the probability density function to obtain an interval joint probability density function. However, we will do something more general and model the uncertainty in the θ_i by substituting a fuzzy number for θ_i and obtain a joint fuzzy probability density function. In fact one could think of confidence intervals for a θ_i as making up the α-cuts of the fuzzy number we use for θ_i. See Section 2.8. Using α-cuts of the fuzzy numbers produces the interval probability density functions. Substituting fuzzy numbers for the uncertain parameters produces joint fuzzy density functions.

In the next subsection we study fuzzy marginals, followed by fuzzy conditionals and fuzzy correlation. Then in subsection 10.2.4 we look at the fuzzy bivariate normal density.

10.2.1   Fuzzy Marginals

X and Y are random variables with joint density $f(x, y; \theta)$. Since the θ_i in θ are uncertain we substitute fuzzy numbers $\overline{\theta}_i$ for the θ_i, $1 \leq i \leq n$, giving joint fuzzy density $f(x, y; \overline{\theta})$. Computing fuzzy probabilities from fuzzy densities was discussed in Chapter 8. The fuzzy marginal for X is

$$f(x; \overline{\theta}) = \int_{-\infty}^{\infty} f(x, y; \overline{\theta}) dy. \tag{10.1}$$

A similar equation for the fuzzy marginal for Y. We compute the α-cuts of the fuzzy marginals $f(x; \overline{\theta})$ as follows

$$f(x; \overline{\theta})[\alpha] = \{ \int_{-\infty}^{\infty} f(x, y; \theta) dy | \theta_i \in \overline{\theta}_i[\alpha], 1 \leq i \leq n \}, \tag{10.2}$$

for $0 \leq \alpha \leq 1$, and a similar equation for $f(y; \overline{\theta})[\alpha]$. Equation (10.2) gives the α-cuts of a fuzzy set for each value of x.

Now suppose $f(x; \theta)$ is the crisp (not fuzzy) marginal of x. We use $f(x; \theta)$ to find the mean $\mu_x(\theta)$ and variance $Var_x(\theta)$ of X. The mean and variance of X may depend on the values of the parameters, so they are written as functions of θ. We assume that $\mu_x(\theta)$ and $Var_x(\theta)$ are continuous functions of θ. We fuzzify the crisp mean and variance to obtain the fuzzy mean and variance of the fuzzy marginal. The following theorem is for X but a similar one holds for Y.

Theorem1. The fuzzy mean and variance of the fuzzy marginal $f(x; \overline{\theta})$ are $\mu_x(\overline{\theta})$ and $Var_x(\overline{\theta})$.

Proof : An α-cut of the fuzzy mean of the fuzzy marginal for X is

$$\mu_x(\overline{\theta})[\alpha] = \{ \int_{-\infty}^{\infty} x f(x; \theta) dx | \theta_i \in \overline{\theta}_i[\alpha], 1 \leq i \leq n \}, \tag{10.3}$$

for $0 \leq \alpha \leq 1$. Now the integral in equation (10.3) equals $\mu_x(\theta)$ for each $\theta_i \in \overline{\theta}_i$, $1 \leq i \leq n$. So

$$\mu_x(\overline{\theta})[\alpha] = \{ \mu_x(\theta) | \theta_i \in \overline{\theta}_i[\alpha], 1 \leq i \leq n \}. \tag{10.4}$$

Hence, the fuzzy mean is $\mu_x(\overline{\theta})$. See the "applications" part of Section 2.4.

The α-cuts of the fuzzy variance are

$$Var_x(\overline{\theta})[\alpha] = \{ \int_{-\infty}^{\infty} (x - \mu_x(\theta))^2 f(x; \theta) dx | \theta_i \in \overline{\theta}_i[\alpha], 1 \leq i \leq n \}, \tag{10.5}$$

for $0 \leq \alpha \leq 1$. But the integral in the above equation equals $Var_x(\theta)$ for each $\theta_i \in \overline{\theta}_i[\alpha]$, $1 \leq i \leq n$. Hence

$$Var_x(\overline{\theta})[\alpha] = \{ Var_x(\theta) | \theta_i \in \overline{\theta}_i[\alpha], 1 \leq i \leq n \}. \tag{10.6}$$

So, the fuzzy variance is just $Var_x(\overline{\theta})$.

Example 10.2.1.1

This example will be continued through the next two subsections. Let the joint density be $f(x, y; \lambda) = \lambda^2 e^{-\lambda y}$ for $0 < x < y$, and zero otherwise. The parameter $\lambda > 0$. The marginals are $f(x; \lambda) = \lambda e^{-\lambda x}$, $x > 0$ and $f(y; \lambda) = \lambda^2 y e^{-\lambda y}$ for $y > 0$. From this we find $\mu_x(\lambda) = \lambda^{-1}$, $\mu_y(\lambda) = 2/\lambda$, $Var_x(\lambda) = \lambda^{-2}$ and $Var_y(\lambda) = 2/\lambda^2$. Now consider the joint fuzzy density.

Let $\overline{\lambda} > 0$ be a fuzzy number. The joint fuzzy density is $f(x, y; \overline{\lambda}) = \overline{\lambda}^2 e^{-\overline{\lambda} y}$ for $0 < x < y$. The fuzzy marginal for X is

$$f(x; \overline{\lambda}) = \int_x^\infty \overline{\lambda}^2 e^{-\overline{\lambda} y} dy. \tag{10.7}$$

This fuzzy marginal has α-cuts

$$f(x; \overline{\lambda})[\alpha] = \{ \int_x^\infty \lambda^2 e^{-\lambda y} dy | \lambda \in \overline{\lambda}[\alpha] \}, \tag{10.8}$$

for $0 \le \alpha \le 1$. The integral in equation (10.8) equals $\lambda e^{-\lambda x}$. So

$$f(x; \overline{\lambda})[\alpha] = \{ \lambda e^{-\lambda x} | \lambda \in \overline{\lambda}[\alpha] \}, \tag{10.9}$$

for $0 \le \alpha \le 1$. Although this looks like we get a simple result for the fuzzy marginal for X, we do not. First let $g(\lambda) = \lambda e^{-\lambda x}$ be a function of only λ for a fixed value of $x > 0$ and let $\lambda[0] = [\lambda_1, \lambda_2]$. Now $g(\lambda)$ is an increasing function of λ for $\lambda < 1/x$ and it is a decreasing function of λ for $\lambda > 1/x$. Let $f(x; \overline{\lambda})[\alpha] = [f_1(x, \alpha), f_2(x, \alpha)]$. To find the end points of these α-cuts we must solve

$$f_1(x, \alpha) = min\{ g(\lambda) | \lambda \in \overline{\lambda}[\alpha] \}, \tag{10.10}$$

and

$$f_2(x, \alpha) = max\{ g(\lambda) | \lambda \in \overline{\lambda}[\alpha] \}, \tag{10.11}$$

for $0 \le \alpha \le 1$ and for $x > 0$. We see the solutions to equations (10.10) and (10.11) will depend on whether $1/x < \lambda_1$, or $\lambda_1 \le 1/x \le \lambda_2$ or $\lambda_2 < 1/x$. So we obtain fairly complicated fuzzy sets for $f(x; \overline{\lambda})$ as x increases from zero to ∞. However, in the calculations we wish to do we will not need to determine all these fuzzy sets precisely.

The fuzzy mean of X, from Theorem 1, is $1/\overline{\lambda}$, and the fuzzy variance of X is $1/\overline{\lambda}^2$. Let us find the fuzzy probability $\overline{A}$ that X takes on a value between a and b for $0 < a < b$. This fuzzy probability is

$$\overline{A} = \int_a^b f(x; \overline{\lambda}) dx. \tag{10.12}$$

The integral is evaluated using α-cuts as

$$\overline{A}[\alpha] = \{ \int_a^b f(x; \lambda) dx | \lambda \in \overline{\lambda}[\alpha] \}, \tag{10.13}$$

which equals

$$\overline{A}[\alpha] = \{\int_a^b \lambda e^{-\lambda x}dx | \lambda \in \overline{\lambda}[\alpha]\}, \tag{10.14}$$

which, after evaluating the integral, is

$$\overline{A}[\alpha] = \{e^{-a\lambda} - e^{-b\lambda} | \lambda \in \overline{\lambda}[\alpha]\}, \tag{10.15}$$

for $0 \leq \alpha \leq 1$. From this last equation we may find the end points of the intervals and construct the fuzzy probability $\overline{A}$. Sometimes finding these end points is computationally difficult, and then we would consider employing an evolutionary algorithm or look at the methods described in Section 2.9.

The fuzzy marginal for Y is

$$f(y; \overline{\lambda}) = \int_0^y \overline{\lambda}^2 e^{-\overline{\lambda}y}dx. \tag{10.16}$$

This integral is also evaluated through α-cuts

$$f(y; \overline{\lambda})[\alpha] = \{\int_0^y \lambda^2 e^{-\lambda}dx | \lambda \in \overline{\lambda}[\alpha]\}, \tag{10.17}$$

which equals

$$f(y; \overline{\lambda})[\alpha] = \{\lambda^2 y e^{-\lambda y} | \lambda \in \overline{\lambda}[\alpha]\}, \tag{10.18}$$

for all α.

The fuzzy mean of Y, from Theorem 1, is $2/\overline{\lambda}$ and the fuzzy variance of Y is $2/\overline{\lambda}^2$. If $\overline{B} = Prob[a < Y < b]$, then

$$\overline{B} = \int_a^b f(y, \overline{\lambda})dy. \tag{10.19}$$

Alpha-cuts of $\overline{B}$ are

$$\overline{B}[\alpha] = \{\int_a^b \lambda^2 y e^{-\lambda y}dy | \lambda \in \overline{\lambda}[\alpha]\}, \tag{10.20}$$

for $\alpha \in [0, 1]$. We evaluate the integral in the last equation and then compute the end points of the intervals to obtain the fuzzy probability $\overline{B}$.

10.2.2   Fuzzy Conditionals

The conditional density of X given that $Y = y$ is

$$f(x|y; \theta) = \frac{f(x, y; \theta)}{f(y; \theta)}, \tag{10.21}$$

provided that $f(y; \theta) \neq 0$, and the conditional density of Y given that $X = x$ is

$$f(y|x; \theta) = \frac{f(x, y; \theta)}{f(x; \theta)}, \tag{10.22}$$

given that $f(x; \theta) \neq 0$. From these conditionals we may find the conditional mean of X (Y) given $Y = y$ ($X = x$) written as $\mu_{x|y}(\theta)$ ($\mu_{y|x}(\theta)$) and we can compute the conditional variance of X (Y) given $Y = y$ ($X = x$) written as $Var_{x|y}(\theta)$ ($Var_{y|x}(\theta)$).

Now we proceed to the fuzzy case. Alpha-cuts of $f(x|y; \bar{\theta})$ are

$$f(x|y; \bar{\theta})[\alpha] = \left\{ \frac{f(x, y; \theta)}{f(y; \theta)} | \theta_i \in \bar{\theta}_i[\alpha], 1 \leq i \leq n \right\}, \qquad (10.23)$$

for all α. Similarly we define $f(y|x; \bar{\theta})[\alpha]$. Then

$$\mu_{x|y}(\bar{\theta}) = \int_{-\infty}^{\infty} x f(x|y; \bar{\theta}) dx, \qquad (10.24)$$

whose α-cuts are

$$\mu_{x|y}(\bar{\theta})[\alpha] = \{ \int_{-\infty}^{\infty} x \frac{f(x, y; \theta)}{f(y; \theta)} dx | \theta_i \in \bar{\theta}_i[\alpha], 1 \leq i \leq n \}, \qquad (10.25)$$

for $0 \leq \alpha \leq 1$. Similarly we may specify the other fuzzy conditional mean and the fuzzy conditional variances.

Theorem 2 The fuzzy mean of X (Y) given $Y = y$ ($X = x$) is $\mu_{x|y}(\bar{\theta})$ ($\mu_{y|x}(\bar{\theta})$) and the fuzzy variance of X (Y) given $Y = y$ ($X = x$) is $Var_{x|y}(\bar{\theta})$ ($Var_{y|x}(\bar{\theta})$).

Proof : Similar to the proof of Theorem 1.

Example 10.2.2.1

This continues Example 10.3.1.1. We determine $f(x|y; \lambda) = 1/y$, for $0 < x < y$, and $f(y|x; \lambda) = (\lambda e^{\lambda x}) e^{-\lambda y}$, for $0 < x < y$, and $\mu_{x|y}(\lambda) = y/2$, $\mu_{y|x}(\lambda) = x + \frac{1}{\lambda}$, and from these results we may also find $Var_{x|y}(\lambda)$ and $Var_{y|x}(\lambda)$. So from Theorem 2 we may find the fuzzy conditional means and variances.

Let us go through some of these calculations for the fuzzy conditional mean. We will concentrate on the fuzzy conditional density of X given $Y = y$. By definition

$$f(x|y; \bar{\lambda}) = \frac{f(x, y, \bar{\lambda})}{f(y; \bar{\lambda})}, \qquad (10.26)$$

whose α-cuts are

$$f(x|y; \bar{\lambda})[\alpha] = \left\{ \frac{f(x, y; \lambda)}{f(y; \lambda)} | \lambda \in \bar{\lambda}[\alpha] \right\}, \qquad (10.27)$$

which equals

$$f(x|y; \bar{\lambda})[\alpha] = \left\{ \frac{1}{y} | \lambda \in \bar{\lambda}[\alpha] \right\}, \qquad (10.28)$$

for $0 \leq \lambda \leq 1$, for $0 < x < y$. Hence, this fuzzy conditional density is crisp and

$$f(x|y, \overline{\lambda}) = \frac{1}{y}, 0 < x < y. \tag{10.29}$$

Then

$$\mu_{x|y}(\overline{\lambda}) = \int_{-\infty}^{\infty} x f(x|y; \overline{\lambda}) dx, \tag{10.30}$$

whose α-cuts are

$$\mu_{x|y}(\overline{\lambda})[\alpha] = \{ \int_0^y x f(x|y; \lambda) dx | \lambda \in \overline{\lambda}[\alpha] \}, \tag{10.31}$$

which equals

$$\mu_{x|y}(\overline{\lambda})[\alpha] = \{ \int_0^y \frac{x}{y} dx | \lambda \in \overline{\lambda}[\alpha] \}, \tag{10.32}$$

and so we obtain

$$\mu_{x|y}(\overline{\lambda}) = \frac{y}{2}, \tag{10.33}$$

a crisp result.

To conclude this example we now find the fuzzy probability $\overline{C} = Prob[a < Y < b | X = c]$, assuming $f(c; \overline{\lambda}) \neq 0$, where $c < a < b$, $0 < c < y$. The basic equation for this fuzzy probability is

$$\overline{C} = \int_a^b f(y|c; \overline{\lambda}) dy, \tag{10.34}$$

whose α-cuts are

$$\overline{C}[\alpha] = \{ \int_a^b f(y|c; \lambda) dy | \lambda \in \overline{\lambda}[\alpha] \}. \tag{10.35}$$

We substitute for $f(y|c; \lambda)$ and evaluate the resulting integral producing the following α-cuts for $\overline{C}$

$$\overline{C}[\alpha] = \left\{ \lambda e^{\lambda c} \left[\frac{e^{-\lambda b} - e^{-\lambda a}}{\lambda} \right] | \lambda \in \overline{\lambda}[\alpha] \right\}. \tag{10.36}$$

10.2.3   Fuzzy Correlation

The covariance of X and Y is

$$Cov(x, y; \theta) = \mu_{xy}(\theta) - \mu_x(\theta)\mu_y(\theta), \tag{10.37}$$

where

$$\mu_{xy}(\theta) = \int_{-\infty}^{\infty} \int_{-\infty}^{\infty} xy f(x, y; \theta) dx dy. \tag{10.38}$$

Then the correlation coefficient between X and Y is

$$\rho_{xy}(\theta) = \frac{Cov(x,y;\theta)}{\sqrt{Var_x(\theta)Var_y(\theta)}}. \tag{10.39}$$

Now we want to find the fuzzy correlation between X and Y. All the items in the fuzzy correlation have been previously defined except the fuzzy covariance. So, we need to specify $\mu_{xy}(\overline{\theta})$. We first find the α-cuts of $\mu_{xy}(\overline{\theta})$

$$\mu_{xy}(\overline{\theta})[\alpha] = \{\int_{-\infty}^{\infty}\int_{-\infty}^{\infty} xyf(x,y;\theta)dxdy | \theta_i \in \overline{\theta}_i[\alpha], 1 \leq i \leq n\}. \tag{10.40}$$

Using this result we may find $Cov(x,y;\overline{\theta})$ and then the fuzzy correlation $\rho_{xy}(\overline{\theta})$.

Example 10.2.3.1

This continues Examples 10.2.1.1 and 10.2.2.1. All we need is the calculation $\mu_{xy}(\lambda) = 3/\lambda^2$. Then, as in Theorems 1 and 2, we see that $\mu_{xy}(\overline{\lambda}) = 3/\overline{\lambda}^3$. The fuzzy correlation is

$$\rho_{xy}(\overline{\lambda}) = \frac{\mu_{xy}(\overline{\lambda}) - \mu_x(\overline{\lambda})\mu_y(\overline{\lambda})}{\sqrt{Var_x(\overline{\lambda})Var_y(\overline{\lambda})}}. \tag{10.41}$$

The fuzzy correlation is evaluated using α-cuts. So

$$\rho_{xy}(\overline{\lambda})[\alpha] = \left\{ \frac{\frac{3}{\lambda^2} - \frac{1}{\lambda}\frac{2}{\lambda}}{\sqrt{\frac{1}{\lambda^2}\frac{2}{\lambda^2}}} | \lambda \in \overline{\lambda}[\alpha] \right\}, \tag{10.42}$$

which simplifies to simply $\frac{1}{\sqrt{2}}$ for all α (same as the crisp result). Hence, in this example, the fuzzy correlation turns out to be a crisp number.

10.2.4 Fuzzy Bivariate Normal

The bivariate normal is defined as [2]

$$f(x,y;\theta) = Ke^{-Q/2}, \tag{10.43}$$

where

$$\theta = (\mu_x, \mu_y, \sigma_x, \sigma_y, \rho), \tag{10.44}$$

for $-\infty < \mu_x, \mu_y < \infty$, $0 < \sigma_x, \sigma_y$, $-1 < \rho < 1$, and

$$K = K(\theta_1) = [2\pi\sigma_x\sigma_y\sqrt{1-\rho^2}]^{-1}, \tag{10.45}$$

for $\theta_1 = (\sigma_x, \sigma_y, \rho)$, and

$$Q = Q(\theta) = \frac{1}{1 - \rho^2}[(\frac{x - \mu_x}{\sigma_x})^2 - 2\rho(\frac{x - \mu_x}{\sigma_x})(\frac{y - \mu_y}{\sigma_y}) + (\frac{y - \mu_y}{\sigma_y})^2]. \quad (10.46)$$

We will now use the notation $N(a, b)$ for the normal probability density with mean a and variance b. We know that the marginal for X is $f(x; \theta_2) = N(\mu_x, \sigma_x)$, $\theta_2 = (\mu_x, \sigma_x)$ and the marginal for Y is $f(y; \theta_3) = N(\mu_y, \sigma_y)$, $\theta_3 = (\mu_y, \sigma_y)$. Also, ρ is the correlation between X and Y. The conditional density for X given $Y = y$ is $f(x|y; \theta) = N(\mu_{x|y}(\theta), Var_{x|y}(\theta_4))$, where $\mu_{x|y}(\theta) = \mu_x + \rho \frac{\sigma_x}{\sigma_y}(y - \mu_y)$ and $Var_{x|y}(\theta_4) = \sigma_x^2(1 - \rho^2)$, $\theta_4 = (\sigma_x, \rho)$. There is a similar expression for the conditional density of Y given $X = x$ which we shall omit.

Now we substitute fuzzy number $\overline{\mu}_x$, $\overline{\mu}_y$, ... for μ_x, μ_y,... respectively to obtain the fuzzy bivariate normal $f(x, y; \overline{\theta})$. In what follows the notation $\theta \in \overline{\theta}[\alpha]$ means that $\theta_i \in \overline{\theta}_i[\alpha]$, $1 \le i \le 5$. Similarly, we shall interpret $\theta_j \in \overline{\theta}_j[\alpha]$, $j = 1, 2, 3, 4$.

The fuzzy marginal for X is

$$f(x; \overline{\theta}_2) = \int_{-\infty}^{\infty} f(x, y; \overline{\theta})dy, \quad (10.47)$$

whose α-cuts are

$$f(x; \overline{\theta}_2) = \{\int_{-\infty}^{\infty} f(x, y; \theta)dy | \theta \in \overline{\theta}[\alpha]\}, \quad (10.48)$$

which equals

$$\{N(\mu_x, \sigma_x^2) | \mu_x \in \overline{\mu}_x[\alpha], \sigma_x^2 \in \overline{\sigma}_x^2[\alpha]\}. \quad (10.49)$$

So, $f(x; \overline{\theta}_2) = N(\overline{\mu}_x, \overline{\sigma}_x^2)$. See the "applications" part of Section 2.4. Similarly we get the fuzzy marginal for Y to be $f(y, \overline{\theta}_3) = N(\overline{\mu}_y, \overline{\sigma}_y^2)$.

Suppose we wish to find the fuzzy probability $\overline{D}$ that X takes its values in the interval $[c, d]$. The α-cuts of this fuzzy probability are

$$\overline{D}[\alpha] = \{\int_c^d N(\mu_x, \sigma_x^2)dx | \mu_x \in \overline{\mu}_x[\alpha], \sigma_x^2 \in \overline{\sigma}_x^2[\alpha]\}. \quad (10.50)$$

Now we consider the fuzzy conditional densities and we will only look at $f(x|y; \overline{\theta})$ whose α-cuts are

$$f(x|y; \overline{\theta})[\alpha] = \left\{ \frac{f(x, y; \theta)}{f(y; \theta_3)} | \theta \in \overline{\theta}[\alpha] \right\}, \quad (10.51)$$

which equals

$$\{N(\mu_{x|y}(\theta), Var_{x|y}(\theta_4)) | \theta \in \overline{\theta}[\alpha]\}. \quad (10.52)$$

Hence

$$f(x|y; \overline{\theta}) = N(\mu_{x|y}(\overline{\theta}), Var_{x|y}(\overline{\theta}_4)). \quad (10.53)$$

For example, the fuzzy mean in the above equation is evaluated as

$$\mu_{x|y}(\bar{\theta})[\alpha] = \{\mu_{x|y}(\theta) | \theta \in \bar{\theta}[\alpha]\}. \tag{10.54}$$

If $\bar{E}$ is the fuzzy probability that X takes a value in the interval $[e, f]$, given $Y = c$, then α-cuts of this fuzzy probability are

$$\bar{E}[\alpha] = \{\int_e^f N(\mu_{x|c}(\theta), Var_{x|c}(\theta_4))dx | \theta \in \bar{\theta}[\alpha]\}. \tag{10.55}$$

10.3 References

1. J.J.Buckley: Uncertain Probabilities III: The Continuous Case. Under review.

2. R.V.Hogg and A.T.Craig, Introduction to Mathematical Statistics, Fourth Edition, Macmillan, N.Y., N.Y., 1970.

Chapter 11

Applications of Joint Distributions

11.1   Introduction

This Chapter presents two applications of joint fuzzy probability distributions discussed in Chapter 10. In the next section we have an application of a joint fuzzy discrete probability distribution followed by a discussion of fuzzy reliability theory which is an application of joint fuzzy continuous probability distribution.

11.2   Political Polls

Suppose in a certain midwestern state in the US, which we shall call state W, the results of a recent random sample of 1000 possible voters in the next presidential election was: (1) $p_1 = 0.46$, or 46% said they would vote republican; (2) $p_2 = 0.42$, or 42% said they would vote democratic; and (3) $p_3 = 0.12$, or 12% said they were undecided. The margin of error in this poll was plus, or minus, three percentage points. Let n be the total number of possible voters in state W in the next presidential election. If $X =$ the total numbers that will vote republican and set $Y =$ all those that will vote democratic, then define $f(x, y; p)$ to be the probability that $X = x$, $Y = y$ and $p = (p_1, p_2) = (0.46, 0.42)$ using the point estimates for the p_i. Then we have the discrete trinomial probability mass function

$$f(x, y; p) = \frac{n!}{x!y!z!}(0.46)^x(0.42)^y(0.12)^z, \tag{11.1}$$

where $z = n - x - y$. However, the estimates are uncertain and we now substitute fuzzy numbers for the p_i as follows: (1) $\bar{p}_1 = (0.43/0.46/0.49)$ for

p_1; (2) $\bar{p}_2 = (0.39/0.42/0.45)$ for p_2; and (3) $\bar{p}_3 = (0.06/0.12/0.18)$ for p_3. We doubled the length of the support of $\bar{p}_3$ for computational convenience in Subsection 11.2.2. Now we have a fuzzy trinomial probability mass function

$$f(x, y; \bar{p}) = \frac{n!}{x!y!z!}\overline{p}_1^x\overline{p}_2^y\overline{p}_3^z. \tag{11.2}$$

We will now consider the fuzzy marginals, the fuzzy conditional distributions and fuzzy correlation.

11.2.1 Fuzzy Marginals

Alpha-cuts of the fuzzy marginal for X are

$$f(x; \bar{p})[\alpha] = \{\sum_{y=0}^{n} \frac{n!}{x!y!z!}p_1^x p_2^y p_3^z | \ \mathbf{S} \ \}, \tag{11.3}$$

where $z = n - x - y$ and $\mathbf{S}$ is the statement " $p_i \in \bar{p}_i[\alpha]$, $1 \le i \le 3$, and $p_1 + p_2 + p_3 = 1$". But we know that [2] the crisp marginal for X is $b(n, p_1)$, or the binomial with probability p_1. So equation (11.3) simplifies to

$$f(x, \bar{p})[\alpha] = \{\frac{n!}{x!(n-x)!}p_1^x(1 - p_1)^{(n-x)} | \ \mathbf{S} \ \}. \tag{11.4}$$

Hence $f(x; \bar{p}) = b(n, \bar{p}_1)$, the fuzzy binomial. In equation (11.4) all values of p_1 in $\bar{p}_1[\alpha]$ are feasible because given any value of $p_1 \in \bar{p}_1[\alpha]$ we may find a value of $p_2 \in \bar{p}_2[\alpha]$ and a value of $p_3 \in \bar{p}_3[\alpha]$ so that $p_1 + p_2 + p_3 = 1$. Therefore, see Section 4.2, $\mu_x(\bar{p}) = n\bar{p}_1$ and $Var_x(\bar{p}) \le n\bar{p}_1(1 - \bar{p}_1)$. We obtain similar results for the marginal for Y.

Example 11.2.1.1

Assume the values of the $\bar{p}_i$ given at the beginning of this section, $1 \le i \le 3$. Then

$$\mu_x(\bar{p}) = (0.43n/0.46n/0.49n), \tag{11.5}$$

a triangular fuzzy number. In Example 4.2.2 in Chapter 4 we discussed finding the α-cuts of the fuzzy variance for the fuzzy binomial when $q = 1 - p$. This is the case we have here. But $p_1 = 0.5$ does not belong to any α-cut of $\bar{p}_1$. Hence

$$Var_x(\bar{p})[\alpha] = [np_{11}(\alpha)(1 - p_{11}(\alpha)), np_{12}(\alpha)(1 - p_{12}(\alpha))], \tag{11.6}$$

where $\bar{p}_1[\alpha] = [p_{11}(\alpha), p_{12}(\alpha)] = [0.43 + 0.03\alpha, 0.49 - 0.03\alpha]$. The graph of the fuzzy variance, using $n = 1000$, is given in Figure 11.1.

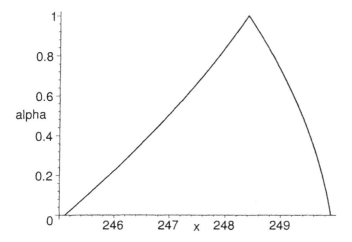

Figure 11.1: Fuzzy Variance in Example 11.2.1.1

11.2.2 Fuzzy Conditionals

We denote the fuzzy conditional of X given $Y = y$ as $f(x|y; \bar{p})$. Computations for the fuzzy conditional of Y given $X = x$ as similar and omitted. Its α-cuts are

$$f(x|y; \bar{p})[\alpha] = \{ \frac{f(x, y; \bar{p})}{f(y; \bar{p})} |\ \ \mathbf{S}\ \ \}. \tag{11.7}$$

In the crisp case we know [2] that this conditional is $b(m, u)$ for $m = n - y$ and $u = \frac{p_1}{1-p_2}$. So equation (11.7) simplifies to

$$f(x|y; \bar{p})[\alpha] = \{ \frac{m!}{x!(m-x)!} u^x (1-u)^{m-x} |\ \ \mathbf{S}\ \ \}. \tag{11.8}$$

Equation (11.8) gives the α-cuts of a fuzzy binomial $b(m, \bar{u})$ and we now need to determine $\bar{u}$. The fuzzy probability $\bar{u}$ has α-cuts

$$\bar{u}[\alpha] = \{ \frac{p_1}{1 - p_2} |\ \ \mathbf{S}\ \ \}. \tag{11.9}$$

Let $\bar{p}_i[\alpha] = [p_{i1}(\alpha), p_{i2}(\alpha)]$ for $1 \leq i \leq 3$. It is easy to see that the expression $\frac{p_1}{1-p_2}$ is an increasing function of both p_1 and p_2. So,

$$\bar{u}[\alpha] = [\frac{p_{11}(\alpha)}{1 - p_{21}(\alpha))}, \frac{p_{12}(\alpha)}{1 - p_{22}(\alpha))}], \tag{11.10}$$

for $0 \leq \alpha \leq 1$. We doubled the support of $\bar{p}_3$ so that now, for any α-cut, p_1 and p_2 can range throughout the α-cuts of $\bar{p}_1$ and $\bar{p}_2$, respectively.

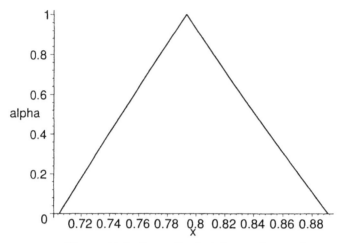

Figure 11.2: Fuzzy Probability in Example 11.2.2.1

From these results we may now compute the conditional mean and variance of X given $Y = y$. Since $f(x|y; \overline{p})$ is a fuzzy binomial (see Section 4.2), we get $\mu_{x|y}(\overline{p}) = m\overline{u}$ and $Var_{x|y}(\overline{p}) \leq m\overline{u}(1 - \overline{u}))$, where $m = n - y$.

Example 11.2.2.1

This example continues Example 11.2.1.1. Putting in the values of the $\overline{p}_i$ and their α-cuts we may find the fuzzy probability $\overline{u}$ whose graph is now in Figure 11.2. The fuzzy number in Figure 11.2 may look like a triangular fuzzy number but it is not a perfect triangle; the sides are slightly curved.

Then $\mu_{x|y}(\overline{p})$ is just $n - y$ times $\overline{u}$. It remains to find the fuzzy conditional variance. Let $\overline{u}[\alpha] = [u_1(\alpha), u_2(\alpha)]$. We see that 0.5 lies to the left of all the α-cuts of $\overline{u}$. Therefore, see Section 4.2, we have

$$Var_{x|y}(\overline{p})[\alpha] = [nu_2(\alpha)(1 - u_2(\alpha)), nu_1(\alpha)(1 - u_1(\alpha))], \qquad (11.11)$$

for all α. The graph of the fuzzy conditional variance, using $n = 1000$, is displayed in Figure 11.3.

11.2.3 Fuzzy Correlation

We may set up the equations for fuzzy correlation as equation (10.39) in Chapter 10, however it is known that in the crisp case [2] we get

$$\rho = -\sqrt{\frac{p_1 p_2}{(1 - p_1)(1 - p_2)}}. \qquad (11.12)$$

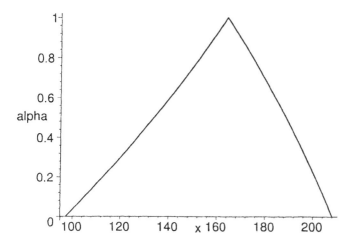

Figure 11.3: Fuzzy Conditional Variance in Example 11.2.2.1

Let $G(p_1, p_2)$ be the value of ρ in equation(11.12). So

$$\rho_{xy}(\overline{p})[\alpha] = \{G(p_1, p_2)| \quad \mathbf{S} \quad \}. \tag{11.13}$$

We determine that G is a decreasing function of both p_1 and p_2. Hence

$$\rho_{xy}(\overline{p})[\alpha] = [G(p_{12}(\alpha), p_{22}(\alpha)), G(p_{11}(\alpha), p_{21}(\alpha))], \tag{11.14}$$

for $\alpha \in [0, 1]$.

Example 11.2.3.1

This example continues Examples 11.2.1.1 and 11.2.2.1. To determine $\rho_{xy}(\overline{p})$ we first find the α-cuts of $\overline{p}_1$ and $\overline{p}_2$ and then substitute these into equation (11.14). Having done this we may obtain a picture of the fuzzy correlation which is given in Figure 11.4. We obtain a negative fuzzy number as expected. The sides of the fuzzy number in Figure 11.4 are slightly curved so it is not a triangular fuzzy number.

11.3 Fuzzy Reliability Theory

This application is adapted from some examples and problems in section 7.7 in [2], which was first presented in [1]. We first present the crisp model. Let T be a random variable that measures time to failure for a certain defective electronic device $\mathcal{E}$, and let $f(x; \lambda) = \lambda e^{-\lambda x}$, $x > 0$ and zero otherwise, for

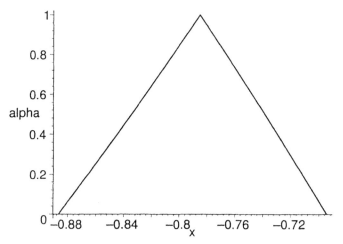

Figure 11.4: Fuzzy Correlation in Example 11.2.3.1

$\lambda > 0$, be the probability density for T. Then $Prob[T > t] = \int_t^\infty f(x; \lambda)dx = e^{-\lambda t}$. Now consider a system $\mathcal{S}$ containing n identical defective copies of $\mathcal{E}$ and let T_i be the time to failure for the ith copy of $\mathcal{E}$, $1 \le i \le n$. We assume all the T_i are independent and the system $\mathcal{S}$ fails if only one of the n defective copies of $\mathcal{E}$ fails. If $M = min\{T_1, ..., T_n\}$, then

$$Prob[M > t] = \prod_{i=1}^{n} Prob[T_i > t] = e^{-n\lambda t}. \qquad (11.15)$$

$Prob[M > t]$ gives the probability that the system $\mathcal{S}$ fails after time t.

Now assume that the system $\mathcal{S}$ has K identical copies of $\mathcal{E}$, some defective and the others non-defective, and each one is defective with probability p independent of the other copies in $\mathcal{E}$. Let N be a random variable giving the number of defective copies of $\mathcal{E}$ in $\mathcal{S}$. So our previous probability calculation $Prob[T > t]$ now becomes a conditional probability $Prob[T > t | N = n] = e^{-n\lambda t}$. The probability distribution for N is binomial

$$Prob[N = n] = \binom{K}{n} p^n q^{K-n}, \qquad (11.16)$$

where $q = 1 - p$. The joint probability that $T > t$ and $N = n$ is

$$Prob[T > t, N = n] = Prob[T > t | N = n]Prob[N = n], \qquad (11.17)$$

which is equal to

$$e^{-n\lambda t} \binom{K}{n} p^n (1 - p)^{K-n}. \qquad (11.18)$$

The unconditional probability $Prob[T > t]$ then is

$$Prob[T > t] = \sum_{n=0}^{K} Prob[T > t, N = n], \quad (11.19)$$

which is

$$\sum_{n=0}^{K} \binom{K}{n} (e^{-\lambda t} p)^n (1 - p)^{K-n}, \quad (11.20)$$

which equals

$$(pe^{-\lambda t} + (1 - p))^K, \quad (11.21)$$

because

$$\sum_{n=0}^{K} \binom{K}{n} a^n b^{K-n} = (a + b)^K. \quad (11.22)$$

This last equation is the final result we wanted to obtain for the crisp model. Now we consider the fuzzy case and show we end up with the fuzzification of equation (11.22). Since the parameter λ in the probability density for T needs to be estimated, and its value is uncertain, we will substitute a fuzzy number $\overline{\lambda}$ for λ and get the fuzzy density $f(x; \overline{\lambda})$ for T. Then the fuzzy probability $Prob[T > t]$ is

$$Prob[T > t] = \int_t^{\infty} \overline{\lambda} e^{-\overline{\lambda} x} dx, \quad (11.23)$$

whose α-cuts are

$$Prob[T > t][\alpha] = \{ \int_t^{\infty} \lambda e^{-\lambda x} dx | \lambda \in \overline{\lambda}[\alpha] \}, \quad (11.24)$$

which is

$$Prob[T > t][\alpha] = \{ e^{-\lambda t} | \lambda \in \overline{\lambda}[\alpha] \}. \quad (11.25)$$

We write this result as $Prob[T > t] = e^{-\overline{\lambda} t}$. Define M as before, the T_i are all independent, then

$$Prob[M > t] = Prob[T_1 > t \quad and....and \quad T_n > t], \quad (11.26)$$

which equals

$$\prod_{i=1}^{n} Prob[T_i > t] = \prod_{i=1}^{n} e^{-\overline{\lambda} t}, \quad (11.27)$$

and the α-cuts of this last expression are

$$\{ \prod_{i=1}^{n} e^{-\lambda t} | \lambda \in \overline{\lambda}[\alpha] \}, \quad (11.28)$$

which equals

$$\{e^{-n\lambda t}|\lambda \in \overline{\lambda}[\alpha]\}. \tag{11.29}$$

So we write $Prob[T > t] = e^{-n\overline{\lambda}t}$. Define K and N as above and the probability p needs to be estimated and is uncertain so we substitute a fuzzy number $\overline{p}$ for p and obtain the fuzzy binomial (see Chapter 4). Alpha-cuts of $Prob[N = n]$ are computed as

$$Prob[N = n][\alpha] = \left\{ \binom{K}{n} p^{n} q^{K-n} | p \in \overline{p}[\alpha], q = 1 - p \right\}. \tag{11.30}$$

We write this probability as

$$Prob[N = n] = \binom{K}{n} \overline{p}^{n} (1 - \overline{p})^{K-n}. \tag{11.31}$$

As before the original probability $Prob[T > t]$ becomes a fuzzy conditional probability $Prob[T > t|N = n]$ and the joint fuzzy probability is

$$Prob[T > t, N = n] = Prob[T > t|N = n]Prob[N = n], \tag{11.32}$$

and then the unconditional probability $Prob[T > t]$ is

$$Prob[T > t] = \sum_{n=0}^{K} e^{-n\overline{\lambda}t} \binom{K}{n} \overline{p}^{n} (1 - \overline{p})^{K-n}, \tag{11.33}$$

whose α-cuts are

$$Prob[T > t][\alpha] = \{ \sum_{n=0}^{K} e^{-n\lambda t} \binom{K}{n} p^{n} q^{K-n} | \lambda \in \overline{\lambda}[\alpha], p \in \overline{p}[\alpha], q = 1 - p \},$$

$$\tag{11.34}$$

which equals

$$\{ (pe^{-\lambda t} + q)^{n} | \lambda \in \overline{\lambda}[\alpha], p \in \overline{p}[\alpha], q = 1 - p \} \tag{11.35}$$

which we write as

$$Prob[T > t] = (\overline{p}e^{-\overline{\lambda}t} + (1 - \overline{p}))^{K}. \tag{11.36}$$

11.4   References

1. J.J.Buckley: Uncertain Probabilities III: The Continuous Case. Under review.

2. R.V.Hogg and E.A.Tanis: Probability and Statistical Inference, Sixth Edition, Prentice Hall, Upper Saddle River, N.J., 2001.

3. D.P.Gaver and G.L.Thompson: Programming and Probability Models in Operations Research, Brooks/Cole, Monterey, Cal., 1973.

Chapter 12

Functions of a Fuzzy Random Variable

12.1 Introduction

In this chapter we will look at a few examples of functions of a single discrete (or continuous) fuzzy random variable. In the next chapter we consider functions of continuous fuzzy random variables (more than one). We start with discrete fuzzy random variables and then we discuss continuous fuzzy random variables.

12.2 Discrete Fuzzy Random Variables

We will consider two elementary examples. The first example is about a simple discrete fuzzy random variable and the second example is a fuzzy random variable having a fuzzy Poisson probability mass function.

Example 12.2.1

Let X be a fuzzy random variable having fuzzy probability mass function $\overline{P}(\{-1\}) = (0.1/0.2/0.3)$, $\overline{P}(\{0\}) = (0.3/0.5/0.70)$ and $\overline{P}(\{1\}) = (0.2/0.3/0.4)$. X takes on only three values $-1, 0, 1$ with the given fuzzy probabilities. If $Y = X^2 - 1$ we want to find the fuzzy probability mass function for the fuzzy random variable Y.

Now Y takes on only two values -1 and 0. Let us determine $\overline{P}(Y = 0)$. Y can equal 0 from $X = -1, 1$. So, the α-cuts of this fuzzy probability are

$$\overline{P}(Y = 0)[\alpha] = \{p_1 + p_3 | \quad \mathbf{S} \quad \}, \tag{12.1}$$

where S is "$p_1 \in \overline{P}(\{-1\})[\alpha]$, $p_2 \in \overline{P}(\{0\})[\alpha]$, $p_3 \in \overline{P}(\{1\})[\alpha]$, and $p_1 + p_2 + p_3 = 1$". In this case we see that $\overline{P}(Y = 0) = (0.3/0.5/0.7)$ because p_1 and p_3 are feasible, see Section 2.9.

Example 12.2.2

Let X be a fuzzy random variable having the fuzzy Poisson (see Section 4.3) $\overline{\lambda}^x \exp(-\overline{\lambda})/x!$ for $x = 0, 1, 2, 3, ...$ and $\overline{\lambda} = (8/10/12)$. If $Y = \sqrt{X}$, find the discrete fuzzy probability mass function for Y. We want $\overline{P}(Y = y)$ for $Y = 0, 1, \sqrt{2}, \sqrt{3}, ...$ Now the function $Y = \sqrt{X}$, or $X = Y^2$, is a one-to-one function for these values of X and Y. It follows that

$$\overline{P}(Y = y) = \overline{\lambda}^{y^2} \exp(-\overline{\lambda})/(y^2)!, \tag{12.2}$$

for $y = 0, 1, \sqrt{2}, \sqrt{3}, ...$

12.3   Continuous Fuzzy Random Variables

We first need to discuss the relationship between probability density functions and distribution functions in both the crisp and the fuzzy cases.

Let X be a random variable having probability density function $f(x; \theta)$ and distribution function $F(x; \theta)$, where $\theta = (\theta_1, ..., \theta_n)$ is a vector of parameters. We know that

$$P(X \in A) = \int_A f(x; \theta)dx, \tag{12.3}$$

and

$$F(x; \theta) = P(X \leq x) = \int_{-\infty}^{x} f(x; \theta)dx, \tag{12.4}$$

and

$$dF(x; \theta)/dx = f(x; \theta). \tag{12.5}$$

Now assume that X is a fuzzy random variable having fuzzy probability density $f(x; \overline{\theta})$ and fuzzy distribution function $F(x; \overline{\theta})$. Let S be the statement "$\theta_i \in \overline{\theta}_i[\alpha]$, $1 \leq i \leq n$". Then, for all $\alpha \in [0, 1]$

$$\overline{P}(X \in A)[\alpha] = \{\int_A f(x; \theta)dx | \ \text{S} \ \}, \tag{12.6}$$

and

$$F(x; \overline{\theta})[\alpha] = \overline{P}(X \leq x)[\alpha] = \{\int_{-\infty}^{x} f(u; \theta)du | \text{S}\}, \tag{12.7}$$

and

$$f(x; \overline{\theta})[\alpha] = \{dF(x; \theta)/dx | \ \text{S} \ \}. \tag{12.8}$$

We will now look at three examples of finding the fuzzy probability density function of a function of a continuous fuzzy random variable. The support of a crisp probability density function $f(x; \theta)$ is $\{x | f(x; \theta) > 0\}$. These examples show that is easy to solve this problem when the support of $f(x; \theta)$ does not depend on θ.

Example 12.3.1

Let X have the fuzzy uniform $U(0, \overline{b})$, $\overline{b} = (0.8/1/1.20)$, see Section 8.2. Let $Y = X^2$ and find the fuzzy distribution function for Y. For $y \geq 0$, let $\overline{P}(Y \leq y)$ denote the fuzzy distribution function. Then $\overline{P}(Y \leq y) = \overline{P}(X^2 \leq y) = \overline{P}(-\sqrt{y} \leq X \leq \sqrt{y})$ whose α-cuts are (see equation (8.1))

$$\{L(-\sqrt{y}, \sqrt{y}, 0, t)/t \mid t \in \overline{b}[\alpha]\}. \tag{12.9}$$

We need to evaluate this equation and then, if possible, equation (12.8) can be used to get the fuzzy probability density function for Y. We will have trouble in applying equation (12.8) because the support of $U(0, b)$ depends on b.

Let $\overline{b}[\alpha] = [b_1(\alpha), b_2(\alpha)]$. First, if $0 \leq \sqrt{y} \leq 0.8$, then we see that

$$\overline{P}(Y \leq y)[\alpha] = [\frac{\sqrt{y}}{b_2(\alpha)}, \frac{\sqrt{y}}{b_1(\alpha)}], \tag{12.10}$$

for all α. Next, if $1.2 \leq \sqrt{y}$, then $\overline{P}(Y \leq y)[\alpha] = [1, 1] = 1$, crisp one, $\alpha \in [0, 1]$. So, now assume that $0.8 < \sqrt{y} \leq 1$. Then

$$\overline{P}(Y \leq y)[\alpha] = [\frac{\sqrt{y}}{b_2(\alpha)}, 1], 0 \leq \alpha \leq \alpha^*, \tag{12.11}$$

and

$$\overline{P}(Y \leq y)[\alpha] = [\frac{\sqrt{y}}{b_2(\alpha)}, \frac{\sqrt{y}}{b_1(\alpha)}], \alpha^* \leq \alpha \leq 1, \tag{12.12}$$

where $b_1(\alpha^*) = \sqrt{y}$. We may also find another expression for the fuzzy distribution function if $1 \leq \sqrt{y} < 1.2$. It is

$$\overline{P}(Y \leq y)[\alpha] = [\frac{\sqrt{y}}{b_2(\alpha)}, 1], 0 \leq \alpha \leq \alpha^*, \tag{12.13}$$

and

$$\overline{P}(Y \leq y)[\alpha] = [1, 1], \alpha^* \leq \alpha \leq 1, \tag{12.14}$$

where $b_2(\alpha^*) = \sqrt{y}$. We may put all this together. Let $\overline{N}(x) = 1$ for $x \leq 1$ and zero otherwise. Then

$$\overline{P}(Y \leq y) = \sqrt{y}(1/\overline{b}) \cap \overline{N}, \tag{12.15}$$

for $0 \leq \sqrt{y} \leq 1.2$. If $\overline{N}$ was not in the expression for the fuzzy distribution function, we could apply equation (12.8) and obtain the fuzzy probability density for Y. We can not use equation (12.8) to find the fuzzy probability density for Y.

Example 12.3.2

Let X be a continuous fuzzy random variable with the fuzzy negative exponential probability density function (see Section 8.4) $f(x; \overline{\lambda}) = \overline{\lambda} \exp(-\overline{\lambda}x)$ for $x \geq 0$ and $f(x; \overline{\lambda}) = 0$ for $x < 0$. If $Y = \sqrt{X}$, then find the fuzzy probability density function for Y. We will find the fuzzy distribution function $G(y; \overline{\lambda})$ for Y and then use equation (12.8) to get the fuzzy probability density function $g(y; \overline{\lambda})$.

We calculate for $y \geq 0$

$$\overline{P}(Y \leq y) = G(y; \overline{\lambda}) = \overline{P}(\sqrt{X} \leq y) = \overline{P}(X \leq y^2), \qquad (12.16)$$

which equals

$$\{ \int_0^{y^2} \lambda \exp(-\lambda x) dx | \lambda \in \overline{\lambda}[\alpha] \}, \qquad (12.17)$$

which is the same as

$$\{ 1 - \exp(-\lambda y^2) | \lambda \in \overline{\lambda}[\alpha] \}. \qquad (12.18)$$

So

$$G(y; \overline{\lambda}) = 1 - \exp(-\overline{\lambda}y^2), \qquad (12.19)$$

for $y \geq 0$ and the fuzzy distribution function is zero otherwise. Hence, by equation (12.8) we obtain

$$g(y; \overline{\lambda}) = 2y\overline{\lambda} \exp(-\overline{\lambda}y^2), \qquad (12.20)$$

for $y \geq 0$ and it is zero otherwise.

Example 12.3.3

X is a continuous fuzzy random variable with the fuzzy normal $N(\overline{\mu}, \overline{\sigma}^2)$ probability density, see Section 8.3. Let $W = \frac{X - \overline{\mu}}{\overline{\sigma}}$ and find the fuzzy probability density function for W. We first specify the α-cuts of $\overline{\sigma}$ from the α-cuts of $\overline{\sigma}^2$. If $\overline{\sigma}^2[\alpha] = [\sigma_1^2(\alpha), \sigma_2^2(\alpha)]$, then $\overline{\sigma}[\alpha] = [\sqrt{\sigma_1^2(\alpha)}, \sqrt{\sigma_2^2(\alpha)}] = [\sigma_1(\alpha), \sigma_2(\alpha)]$.

Let the fuzzy distribution function for W be $H(w; \overline{\mu}, \overline{\sigma})$. We first determine the fuzzy distribution function and use equation (12.8) to find the fuzzy density $h(w; \overline{\mu}, \overline{\sigma})$.

$$\overline{P}(W \leq w) = H(w; \overline{\mu}, \overline{\sigma}) = \overline{P}(\frac{X - \overline{\mu}}{\overline{\sigma}} \leq w), \qquad (12.21)$$

which is interpreted using α-cuts as

$$\overline{P}(W \leq w)[\alpha] = \{P(X \leq \mu + \sigma w | \quad \mathbf{S} \quad \},\tag{12.22}$$

for all α in $[0, 1]$, where $\mathbf{S}$ is "$\mu \in \overline{\mu}[\alpha], \sigma \in \overline{\sigma}[\alpha]$". This last equation is the same as

$$\{ \int_{-\infty}^{\mu + \sigma w} N(\mu, \sigma^2) dx | \quad \mathbf{S} \quad \},\tag{12.23}$$

where $N(\mu, \sigma^2)$ is the normal probability density with mean μ and variance σ^2. Make the change of variable $z = \frac{x - \mu}{\sigma}$ in the integral in the above equation and we have

$$\{ \int_{-\infty}^{w} N(0, 1) dz | \quad \mathbf{S} \quad \}.\tag{12.24}$$

But this last equation is not fuzzy. It equals the crisp distribution function, evaluate at w, for the standard normal ($N(0, 1)$) with mean zero and variance one. Hence W has the crisp $N(0, 1)$ as its probability density function. Does W^2 have the crisp Chi-square, with one degree of freedom, probability density?

Chapter 13

Functions of Fuzzy Random Variables

13.1   Introduction

We first discuss some theoretical results and then look at applications in the next two sections. The theoretical results are stated in terms of continuous fuzzy random variables.

Let $X_1, ..., X_n$ be a random sample from a probability density function $f(x, \theta)$, where $\theta = (\theta_1, ..., \theta_m)$ is a vector of parameters. What this means is that each X_i, $1 \leq i \leq n$, has probability density $f(x; \theta)$ and the X_i, $1 \leq i \leq n$, are independent. Since the X_i, $1 \leq i \leq n$, are independent we may find their joint probability density $f(x_1, ..., x_n; \theta)$ as the product of the individual probability density functions. That is

$$f(x_1, ..., x_n; \theta) = \prod_{i=1}^{n} f(x_i; \theta). \tag{13.1}$$

Now let $X_1, ..., X_n$ be a random sample from the fuzzy probability density $f(x; \overline{\theta})$, where $\overline{\theta} = (\overline{\theta}_1, ..., \overline{\theta}_m)$ is a vector of fuzzy number parameters. What this means is that each X_i, $1 \leq i \leq n$, has fuzzy probability density $f(x; \overline{\theta})$ and the X_i, $1 \leq i \leq n$, are independent. The X_i, $1 \leq i \leq n$, are independent if and only if their joint fuzzy probability density $f(x_1, ...x_n; \overline{\theta})$ is the product of the individual fuzzy probability density functions. That is

$$f(x_1, ..., x_n; \overline{\theta}) = \prod_{i=1}^{n} f(x_i; \overline{\theta}). \tag{13.2}$$

Next we consider transformations of fuzzy random variables. We restrict the discussion, for simplicity, to only two fuzzy random variables. Let X_1

and X_2 be two fuzzy random variables with joint fuzzy probability density $f(x_1, x_2; \overline{\theta})$. Let $\mathcal{A}$ be the support of $f(x_1, x_2; \overline{\theta})$. $\mathcal{A} = \{(x_1, x_2) | f(x_1, x_2; \overline{\theta}) > 0\}$. Next define $Y_1 = u_1(X_1, X_2)$, $Y_2 = u_2(X_1, X_2)$ to be a one-to-one transformation of $\mathcal{A}$ onto the set $\mathcal{B}$ in the $y_1 y_2$-plane. If $A \in \mathcal{A}$, then let B be the mapping of A under this one-to-one transformation. The events $(X_1, X_2) \in A$ and $(Y_1, Y_2) \in B$ are equivalent. Now set $X_1 = v_1(Y_1, Y_2)$, $X_2 = v_2(Y_1, Y_2)$ as the inverse transformation from $\mathcal{B}$ onto $\mathcal{A}$. We next calculate the Jacobian (J) of this inverse transformation. J is the determinant of the 2×2 matrix $[\partial X_i / \partial Y_j]$. We assume that J is not identically zero. Now we can determine the joint fuzzy probability density for Y_1 and Y_2. Call this joint fuzzy probability density $g(y_1, y_2; \overline{\theta})$.

To determine $g(y_1, y_2; \overline{\theta})$ we find the fuzzy probability that $(Y_1, Y_2) \in B$. Suppose that we calculate

$$\overline{P}((Y_1, Y_2) \in B)[\alpha] = \{ \int \int_B h(y_1, y_2; \theta) dy_1 dy_2 | \quad \mathbf{S} \quad \}, \qquad (13.3)$$

for all $\alpha \in [0, 1]$, where $\mathbf{S}$ is "$\theta_i \in \overline{\theta}_i[\alpha]$, $1 \le i \le m$". Then $g(y_1, y_2; \overline{\theta}) = h(y_1, y_2; \overline{\theta})$. Now

$$\overline{P}((Y_1, Y_2) \in B)[\alpha] = \overline{P}((X_1, X_2) \in A)[\alpha], \qquad (13.4)$$

which equals

$$\{ \int \int_A f(x_1, x_2; \theta) dx_1 dx_2 | \quad \mathbf{S} \quad \}, \qquad (13.5)$$

upon change of variables equals

$$\{ \int \int_B f(v_1(y_1, y_2), v_2(y_1, y_2); \theta) |J| dy_1 dy_2 | \quad \mathbf{S} \quad \}. \qquad (13.6)$$

Hence

$$g(y_1, y_2; \overline{\theta}) = f(v_1(y_1, y_2), v_2(y_1, y_2); \overline{\theta}) |J|, \qquad (13.7)$$

for $(y_1, y_2) \in \mathcal{B}$, and is zero otherwise. From the joint fuzzy probability density we may calculate the fuzzy marginals.

The next section contains applications of these results for one-to-one transformations. Section 13.3 has two examples where the transformation is not one-to-one.

13.2   One-to-One Transformation

We will look at three applications of the previous results for one-to-one transformations. Two applications are for continuous fuzzy random variables and one is for discrete fuzzy random variables.

Example 13.2.1

Let X_1, X_2 be a random sample from $N(\overline{\mu}, \overline{\sigma}^2)$. Define $f(x; a, b)$ to be the normal probability density with mean a and variance $b > 0$. The joint fuzzy probability density for X_1, X_2 is $f(x_1, x_2; \overline{\mu}, \overline{\sigma}^2) = f(x_1; \overline{\mu}, \overline{\sigma}^2) f(x_2; \overline{\mu}, \overline{\sigma}^2)$. Let $Y_1 = X_1 + X_2$, $Y_2 = X_1 - X_2$ so that $X_1 = (0.5)(Y_1 + Y_2)$, $X_2 = (0.5)(Y_1 - Y_2)$. The absolute value of the Jacobian equals 0.5. Let the joint fuzzy probability density for Y_1, Y_2 be $g(y_1, y_2; \overline{\mu}, \overline{\sigma}^2)$ and the fuzzy marginal for Y_1 (Y_2) is $g(y_1; \overline{\mu}, \overline{\sigma}^2)$ ($g(y_2; \overline{\mu}, \overline{\sigma}^2)$).

So, from the above discussion, the joint fuzzy probability density for Y_1, Y_2 is

$$(0.5) f(\frac{y_1 + y_2}{2}, \frac{y_1 - y_2}{2}; \overline{\mu}, \overline{\sigma}^2). \tag{13.8}$$

After much algebra, including completing the square, the expression in the above equation simplifies to

$$f(y_1; 2\overline{\mu}, 2\overline{\sigma}^2) f(y_2; 0, 2\overline{\sigma}^2). \tag{13.9}$$

The formula in equation (13.9) gives the joint fuzzy probability density for Y_1, Y_2. We see that Y_1 and Y_2 are independent with the fuzzy marginal for Y_1 (Y_2) $f(y_1; 2\overline{\mu}, 2\overline{\sigma}^2)$ ($f(y_2; 0, 2\overline{\sigma}^2)$).

Example 13.2.2

Let fuzzy random variables X_1, X_2 have joint fuzzy probability density $f(x_1, x_2; \overline{\theta})$. Next let $Y_1 = u_1(X_1, X_2)$. We wish to find the fuzzy probability density for Y_1. What is sometimes done is to let $Y_2 = X_2$ so that we obtain a one-to-one transformation. Find the joint fuzzy probability density for Y_1, Y_2 and then determine the fuzzy marginal for Y_1.

Let X_1, X_2 be a random sample from the fuzzy Poisson (see Section 4.3). The joint fuzzy probability mass function for X_1, X_2 is

$$f(x_1, x_2; \overline{\lambda}) = \overline{\lambda}^{x_1 + x_2} \exp(-2\lambda)/(x_1! x_2!), \tag{13.10}$$

for $x_1, x_2 = 0, 1, 2, 3, \ldots$. Now define $Y_1 = X_1 + X_2$, $Y_2 = X_2$ so that $X_1 = Y_1 - Y_2$, $X_2 = Y_2$. We do not need the Jacobian in the discrete case. Let the joint fuzzy probability mass function for Y_1, Y_2 be $g(y_1, y_2; \overline{\lambda})$. The joint fuzzy probability mass function for Y_1, Y_2 is obtained by substituting $y_1 - y_2$ for x_1, and substituting y_2 for x_2, in equation (13.10). We get

$$g(y_1, y_2; \overline{\lambda}) = \overline{\lambda}^{y_1} \exp(-2\lambda)/(y_1 - y_2)! y_2!, \tag{13.11}$$

for $y_1, y_2 = 0, 1, 2, 3 \ldots$, but $0 \le y_2 \le y_1$. Now we want the fuzzy marginal $g(y_1; \overline{\lambda})$ for Y_1. The α-cuts of this fuzzy marginal are

$$g(y_1; \overline{\lambda})[\alpha] = \{ \sum_{y_2=0}^{y_2=y_1} g(y_1, y_2; \lambda) | \ \mathbf{S} \ \}, \tag{13.12}$$

for all α in $[0, 1]$, where **S** is "$\lambda \in \overline{\lambda}[\alpha]$". We may evaluate the sum in equation (13.12) and obtain

$$g(y_1; \overline{\lambda})[\alpha] = \{(2\lambda)^{y_1} \exp(-2\lambda)/(y_1)! |\quad \mathbf{S}\quad \}. \tag{13.13}$$

Hence, the fuzzy probability density for Y_1 is

$$g(y_1; \overline{\lambda}) = (2\overline{\lambda})^{y_1} \exp(-2\overline{\lambda})/(y_1)!, \tag{13.14}$$

for $y_1 = 0, 1, 2, 3....$ Therefore, Y_1 has the fuzzy Poisson probability density with fuzzy parameter $2\overline{\lambda}$.

Example 13.2.3

Let X_1 and X_2 be a random sample from the fuzzy negative exponential (Section 8.4). The joint fuzzy probability density for X_1, X_2 is

$$f(x_1, x_2; \overline{\lambda}) = \overline{\lambda}^2 \exp(-\overline{\lambda}(x_1 + x_2)), \tag{13.15}$$

for $x_1, x_2 > 0$. Let $Y_1 = X_1/(X_1 + X_2)$, $Y_2 = X_1 + X_2$ so that $X_1 = Y_1 Y_2$, $X_2 = Y_2(1 - Y_1)$. We compute $|J| = y_2$. The support of the joint fuzzy probability density for X_1, X_2 is $\mathcal{A} = \{(x_1, x_2)|x_1 > 0, x_2 > 0\}$ and the transformation of $\mathcal{A}$ is $\mathcal{B} = \{(y_1, y_2)|0 < y_1 < 1, 0 < y_2 < \infty\}$. Let the joint fuzzy probability density for Y_1, Y_2 be $g(y_1, y_2; \overline{\lambda})$ which is obtained by substituting $y_1 y_2$ for x_1, and substituting $y_2(1 - y_1)$ for x_2, in equation (13.15), times $|J|$. We determine

$$g(y_1, y_2; \overline{\lambda}) = \overline{\lambda}^2 y_2 \exp(-\overline{\lambda} y_2), \tag{13.16}$$

for $(y_1, y_2) \in \mathcal{B}$ and zero otherwise. Let $g(y_1; \overline{\lambda})$ and $g(y_2; \overline{\lambda})$ be the fuzzy marginal for Y_1 and Y_2, respectively. We see that Y_1 and Y_2 are independent with $g(y_1; \overline{\lambda})$ the crisp uniform on $[0, 1]$ and

$$g(y_2; \overline{\lambda}) = \overline{\lambda}^2 y_2 \exp(-\overline{\lambda} y_2), \tag{13.17}$$

for $0 < y_2 < \infty$.

13.3   Other Transformations

The two examples in this section have to do with the fuzzy probability densities of two of the order statistics. If $X_1, ..., X_n$ is a random sample from a probability density function $f(x; \theta)$, then, assuming the X_i take on distinct values, $Y_1 < Y_2 < ... < Y_n$ are the order statistics when $Y_1 =$ the smallest of the X_i, $Y_2 =$ the next X_i in order of magnitude,..., and $Y_n =$ the largest X_n. We will be interested in determining the fuzzy probability density functions for Y_1 and Y_n, for $n = 3$, when the random sample is from a fuzzy probability density function. We will be using results from Chapter 12, equations (12.7) and (12.8), in the following examples.

Example 13.3.1

Let X_1, X_2, X_3 be a random sample from the fuzzy negative exponential. If $W = min(X_1, X_2, X_3)$, we want the fuzzy probability density for W. The joint fuzzy probability density for X_1, X_2, X_3 is

$$f(x_1, x_2, x_3; \overline{\lambda}) = \prod_{i=1}^{3} \overline{\lambda} \exp\left(-\overline{\lambda} x_i\right), \tag{13.18}$$

for $x_i > 0$, $1 \le i \le 3$. We denote the fuzzy probability density for W as $g(w; \overline{\lambda})$ and the fuzzy distribution function as $G(w; \overline{\lambda})$. Also define the set $A = \{(x_1, x_2, x_3) | x_i > 0, 1 \le i \le 3, min(x_1, x_2, x_3) \le w\}$ and B is the complement of A so that $B = \{(x_1, x_2, x_3) | x_i > 0 \text{ and } x_i > w, 1 \le i \le 3\}$. Then

$$G(w; \overline{\lambda})[\alpha] = \overline{P}(W \le w)[\alpha] = \{\int \int \int_A f(x_1, x_2, x_3; \lambda) dx_1 dx_2 dx_3 | S\},$$
$$\tag{13.19}$$

for all α in $[0, 1]$, where S is "$\lambda \in \overline{\lambda}[\alpha]$". This last equation equals

$$\{1 - \int \int \int_B f(x_1, x_2, x_3; \lambda) dx_1 dx_2 dx_3 | \quad S \quad \}, \tag{13.20}$$

which is the same as

$$\{1 - \int_w^\infty \int_w^\infty \int_w^\infty f(x_1, x_2, x_3; \lambda) dx_1 dx_2 dx_3 | S\}, \tag{13.21}$$

Finally we get

$$G(w; \overline{\lambda})[\alpha] = \{1 - (e^{-w\lambda})^3 | \quad S \quad \}. \tag{13.22}$$

Hence

$$G(w; \overline{\lambda}) = 1 - e^{-3w\overline{\lambda}}, \tag{13.23}$$

for $w > 0$ and it is zero otherwise. By equation (12.8) we have

$$g(w; \overline{\lambda}) = 3w e^{-3w\overline{\lambda}}, \tag{13.24}$$

for $w > 0$ and zero otherwise.

This example may be generalized to random samples of size $n > 3$ and to other fuzzy probability density functions.

Example 13.3.2

This example continues Example 13.3.1 but now $W = max(X_1, X_2, X_3)$. For $w > 0$ define $A = \{(x_1, x_2, x_3) | 0 < x_i \le w, 1 \le i \le 3\}$. Then

$$G(w, \overline{\lambda})[\alpha] = \{\int \int \int_A f(x_1, x_2, x_3; \lambda) dx_1 dx_2 dx_3 | S\}, \tag{13.25}$$

which equals

$$\left\{ \int_0^w \int_0^w \int_0^w f(x_1, x_2, x_3; \lambda) dx_1 dx_2 dx_3 | \mathbf{S} \right\}, \tag{13.26}$$

or

$$G(w; \overline{\lambda})[\alpha] = \{(1 - e^{-w\lambda})^3 | \mathbf{S}\}. \tag{13.27}$$

Hence

$$G(w; \overline{\lambda}) = (1 - e^{-w\overline{\lambda}})^3, \tag{13.28}$$

for $w > 0$ and zero otherwise. From equation (12.8) we see

$$g(w; \overline{\lambda}) = 3(1 - e^{-w\overline{\lambda}})^2 \overline{\lambda} e^{-w\overline{\lambda}}, \tag{13.29}$$

for $w > 0$.

This example may also be generalized to random samples of size $n > 3$ and to other fuzzy probability density functions.

Chapter 14

Law of Large Numbers

Let $X_1, ..., X_n$ be a random sample from $N(\mu, \sigma^2)$ and define $W = (1/n) \sum_{i=1}^{n} X_i$. We know that W is $N(\mu, \sigma^2/n)$. We need Chebyshev's inequality

$$P(|W - \mu| \geq \epsilon) \leq \frac{\sigma^2}{n\epsilon^2}, \tag{14.1}$$

for any $\epsilon > 0$.

Now assume that $X_1, ..., X_n$ is a random sample from the fuzzy normal probability density $N(\overline{\mu}, \overline{\sigma}^2)$. All we need to assume is that $\overline{\sigma}^2$ is a fuzzy number with bounded support. Again W is the average of the X_i and now W is $N(\overline{\mu}, \overline{\sigma}^2/n)$. We wish to show that

$$limit_{n \to \infty} \overline{P}(|W - \overline{\mu}| \geq \epsilon) = 0, \tag{14.2}$$

which is called the law of large numbers.

First we must define a new method of evaluating $\overline{M} \leq \overline{N}$, where "$\leq$" means "less than or equal to", for two fuzzy numbers $\overline{M}$ and $\overline{N}$, different that studied in Section 2.6. Given two intervals $[a, b]$ and $[c, d]$ we write $[a, b] \leq [c, d]$ if and only if $a \leq c$ and $b \leq d$. Let $\overline{M}[\alpha] = [m_1(\alpha), m_2(\alpha)]$ and $\overline{N}[\alpha] = [n_1(\alpha), n_2(\alpha)]$. Then $\overline{M} \leq \overline{N}$ if and only if $m_1(\alpha) \leq n_1(\alpha)$ and $m_2(\alpha) \leq n_2(\alpha)$ for all $\alpha \in [0, 1]$.

We evaluate the fuzzy probability in equation (14.2) by α-cuts. So

$$\overline{P}(|W - \overline{\mu}| \geq \epsilon)[\alpha] = \{P(|W - \mu| \geq \epsilon)| \quad \mathbf{S} \quad \}, \tag{14.3}$$

for all α where $\mathbf{S}$ is "$\mu \in \overline{\mu}[\alpha]$ and $\sigma^2 \in \overline{\sigma}^2[\alpha]$". From Chebyshev's inequality we see that the interval on right side of the expression in equation (14.3) is less than or equal to the interval

$$\{\sigma^2/(n\epsilon^2)| \quad \mathbf{S} \quad \}, \tag{14.4}$$

for each $\alpha \in [0, 1]$. Therefore

$$\overline{P}(|W - \overline{\mu}| \geq \epsilon)[\alpha] \leq \overline{\sigma}^2[\alpha]/(n\epsilon^2). \tag{14.5}$$

This implies that

$$\overline{P}(|W - \overline{\mu}| \geq \epsilon) \leq \overline{\sigma}^2/(n\epsilon^2). \qquad (14.6)$$

The desired limit result follows.

   This limit may be extended to other probability (mass) density functions.

Chapter 15

Sums of Fuzzy Random Variables

15.1 Introduction

We need to introduce the idea of a fuzzy moment generation function and we will discuss the crisp case first.

Let X be a random variable with probability density $f(x; \theta)$, where θ is a vector of m parameters. We present these results in the section only for the continuous case. Also let $X_1, ..., X_n$ be a random sample from $f(x; \theta)$ and set $Y = X_1 + ... + X_n$. In this chapter the X_i, $1 \leq i \leq n$, will be independent and identically distributed. We will use the symbol "E" for crisp expected value and "$\overline{E}$" for fuzzy expected value. Denote the crisp moment generating function for X as $M_X(t; \theta)$ and it is

$$E(e^{tX}) = \int_{-\infty}^{\infty} e^{tx} f(x; \theta) dx. \tag{15.1}$$

If the moment generating function for Y is $M_Y(t; \theta)$, then we know it is

$$M_Y(t; \theta) = (M_{X_i}(t; \theta))^n. \tag{15.2}$$

There is a uniqueness between probability density functions and moment generating functions. Let Z be a random variable with probability density function $h(z; \tau)$ and moment generating function $M_Z(t; \tau)$. Suppose through other calculations we obtain a random variable T which has the same moment generating function $M_Z(t; \tau)$. Then we can conclude that T has probability density function $h(z; \tau)$.

Another application of the moment generating function is

$$dM_X(0; \theta)/dt = \mu, \tag{15.3}$$

where the expected value of X is μ, and

$$d^2 M_X(0;\theta)/dt^2 - [dM_X(0;\theta)/dt]^2 = \sigma^2, \tag{15.4}$$

where the variance of X is σ^2.

Now assume that X is a fuzzy random variable with fuzzy probability density $f(x;\overline{\theta})$, $X_1, ...X_n$ is a random sample from $f(x;\overline{\theta})$ and set $Y = X_1 + ... + X_n$. We wish to find the fuzzy probability density $g(y,\overline{\theta})$ for Y. Let $M_X(t;\overline{\theta})$ be the fuzzy moment generating function for X and its α-cuts are determined as

$$\overline{E}(e^{tX})[\alpha] = \{ \int_{-\infty}^{\infty} e^{tx} f(x;\theta)dx |\ \ \mathbf{S}\ \ \}, \tag{15.5}$$

for $\alpha \in [0,1]$ and $\mathbf{S}$ is "$\theta_i \in \overline{\theta}_i[\alpha]$, $1 \leq i \leq m$". We find the derivatives of the fuzzy moment generating function through its α-cuts

$$(dM_X(0,\overline{\theta})/dt)[\alpha] = \{dM_X(0;\theta)/dt |\ \ \mathbf{S}\ \ \}, \tag{15.6}$$

for all α in $[0,1]$. We would have a similar expression for the second derivative.

Example 15.1.1

Let the fuzzy random variable X have the fuzzy binomial (Section 4.2)

$$b(n;\overline{p}) = \binom{n}{x} \overline{p}^x (1 - \overline{p})^{n-x}. \tag{15.7}$$

Note that this is different from Section 4.2 where we used $\overline{q}$ in place of $(1-\overline{p})$. Then α-cuts of its fuzzy moment generating function are

$$\overline{E}(e^{tX})[\alpha] = \{\sum_{x=0}^{n} \binom{n}{x} e^{tx} p^x (1 - p)^{n-x} |\ \ \mathbf{S}\ \ \}, \tag{15.8}$$

for α in $[0,1]$ and $\mathbf{S}$ is "$p \in \overline{p}[\alpha]$". But we know, from the moment generating function for the crisp binomial, the the above equation is

$$\{ [(1 - p) + pe^t]^n |\ \ \mathbf{S}\ \ \}, \tag{15.9}$$

so

$$M_X(t,\overline{p}) = [(1 - \overline{p}) + \overline{p}e^t]^n. \tag{15.10}$$

Using the fuzzy moment generating function we find $dM_X(0,\overline{p})/dt = n\overline{p}$. This is the fuzzy mean of the fuzzy binomial when we use $q = 1 - p$, see Section 4.2. Next we can use the fuzzy moment generating function to get the fuzzy variance of X. This computation is a bit more complicated and is done by α-cuts

$$[d^2 M_X(0,\overline{p})/dt^2 - (dM_X(0,\overline{p})/dt)^2][\alpha] = \{np(1 - p)|\mathbf{S}\}, \tag{15.11}$$

which is what we got in Example 4.2.2.

As in the above example we may find the fuzzy moment generating functions for other fuzzy probability (mass) density functions. Let X be a fuzzy random variable having the fuzzy probability mass function the fuzzy Poisson (Section 4.3). The crisp moment generating function for the crisp Poisson is $M_X(t, \lambda) = \exp(\lambda(e^t - 1))$. Hence, the fuzzy moment generating function is $M_X(t; \overline{\lambda}) = \exp(\overline{\lambda}(e^t - 1))$. Next let X be a fuzzy random variable having the fuzzy negative exponential (Section 8.4) as its fuzzy probability density. The crisp moment generating function is $M_X(t; \lambda) = \frac{\lambda}{\lambda - t}$ for $0 < t < \lambda$. So the fuzzy moment generating function is $M_X(t; \overline{\lambda}) = \frac{\overline{\lambda}}{\overline{\lambda} - t}$ for $\overline{\lambda} > 0$, $t > 0$ and t not in the support of $\overline{\lambda}$. Lastly, let X be a fuzzy random variable having as its fuzzy probability density the fuzzy normal $N(\overline{\mu}, \overline{\sigma}^2)$, see Section 8.3. The crisp moment generating function is $M_X(t; \mu, \sigma^2) = \exp(\mu + (t^2/2)\sigma^2)$. It follows that the fuzzy moment generating function is $M_X(t; \overline{\mu}, \overline{\sigma}^2) = \exp(\overline{\mu} + (t^2/2)\overline{\sigma}^2)$.

There is also a uniqueness between fuzzy moment generating functions and fuzzy probability (mass) density functions. Let Z be a fuzzy random variable having fuzzy probability (mass) density function $h(z; \overline{\tau})$ and fuzzy moment generating function $M_Z(t; \overline{\tau})$. Suppose from other calculations we obtain a fuzzy random variable T that has the same fuzzy moment generating function $M_Z(t; \overline{\tau})$. Then we can conclude that T has the fuzzy probability (mass) density function $h(z; \overline{\tau})$.

15.2   Sums

Now we may determine the fuzzy probability (mass) density function for $Y = X_1 + ... + X_n$, where $X_1, ..., X_n$ is a random sample from fuzzy probability (mass) density function $f(x; \overline{\theta})$.

Example 15.2.1

Let X have the fuzzy binomial $b(m; \overline{p})$ defined in equation (15.7) (using m in place of n). Let $M_Y(t; p)$ be the crisp moment generating function for Y. If $M_Y(t; \overline{p})$ is the fuzzy moment generating function for Y its α-cuts are

$$M_Y(t; \overline{p})[\alpha] = \{M_Y(t; p)| \quad \mathbf{S} \quad \}, \tag{15.12}$$

for $\alpha \in [0, 1]$ and $\mathbf{S}$ is "$p \in \overline{p}[\alpha]$". Equation (15.12) equals

$$\{\prod_{i=1}^{n} M_{X_i}(t; p) | \quad \mathbf{S} \quad \}, \tag{15.13}$$

which is the same as

$$\{ [(1-p)+pe^t]^{nm} \mid \quad \mathbf{S} \quad \}. \tag{15.14}$$

Hence

$$M_Y(t; \overline{p}) = [(1 - \overline{p}) + \overline{p}e^t]^{nm}, \tag{15.15}$$

and Y has the fuzzy binomial $b(nm; \overline{p})$ as its fuzzy probability mass function.

Example 15.2.2

Let X have the fuzzy Poisson probability mass function. From the calculations in Example 15.2.1 we see that

$$M_Y(t; \overline{\lambda}) = exp(n\overline{\lambda}(e^t - 1)), \tag{15.16}$$

so Y has the fuzzy Poisson, with parameter $n\overline{\lambda}$, as its fuzzy probability mass function.

Example 15.2.3

Let X be a fuzzy random variable having the fuzzy normal as its fuzzy probability density function. As in Example 15.2.1 we get

$$M_Y(t; \overline{\mu}, \overline{\sigma}^2) = \exp \left(n\overline{\mu} + (t^2/2)n\overline{\sigma}^2 \right), \tag{15.17}$$

and Y has the $N(n\overline{\mu}, n\overline{\sigma}^2)$ as its fuzzy probability density function.

Example 15.2.4

Let X be a fuzzy random variable having the fuzzy negative exponential as its fuzzy probability density function. Then

$$M_Y(t, \overline{\lambda}) = [\overline{\lambda}/(\overline{\lambda} - t)]^n. \tag{15.18}$$

Let $\Gamma(x; \lambda, n)$ denote the gamma probability density function with parameters $\lambda > 0$ and positive integer n. The crisp moment generating function for the gamma is $[\frac{\lambda}{\lambda - t}]^n$. Hence, Y has the fuzzy gamma $\Gamma(x; \overline{\lambda}, n)$ as its fuzzy probability density function.

Chapter 16

Conclusions and Future Research

16.1 Introduction

We first summarize Chapters 3 through 15, without discussing any of the applications contained within a chapter, and then present our suggestions for future research. Our conclusions are at the end of the Chapter.

16.2 Summary

16.2.1 Chapter 3

This chapter introduces our new approach to fuzzy probability. We have a discrete fuzzy probability distribution where $X = \{x_1, ..., x_n\}$, $\overline{P}(\{x_i\}) = \overline{a}_i$, $1 \leq i \leq n$, $\overline{a}_i$ a fuzzy number all i with $0 < \overline{a}_i < 1$ all i. If A and B are subsets of X, we employ restricted fuzzy arithmetic in computing $\overline{P}(A)$, $\overline{P}(A \cup B)$, $\overline{P}(A \cap B)$, etc. We discuss the basic properties of $\overline{P}$ and argue that $\overline{P}(A)$ is also a fuzzy number. Then we go on to fuzzy conditional probability, fuzzy independence and fuzzy Bayes' formula.

16.2.2 Chapter 4

In this chapter we considered two types of discrete fuzzy random variables. We first discussed the fuzzy binomial and then the fuzzy Poisson. In both cases we showed how to find fuzzy probabilities and we computed their fuzzy mean and their fuzzy variance.

16.2.3   Chapter 5

We modeled our fuzzy queuing theory after regular, finite, Markov chains. We showed that we obtain steady state fuzzy probabilities when we use restricted fuzzy arithmetic. We applied these results to the following queuing systems: c parallel, and identical, servers; finite system capacity; and finite, or infinite, calling source.

16.2.4   Chapter 6

We showed that the basic properties of regular, and absorbing, finite Markov chains carry over to fuzzy Markov chains when you use restricted fuzzy arithmetic.

16.2.5   Chapter 7

Here we looked at the classical decision making problem under risk. The probabilities of the states of nature are usually "personal", or subjective, probabilities. Sometimes these probabilities are estimated by "experts". Hence these probabilities are good candidates for fuzzy probabilities. We looked at two cases: (1) fuzzy decisions under risk without data; and (2) fuzzy decisions under risk with data. In the second case we used fuzzy Bayes' formula to update the prior fuzzy probabilities to the posterior fuzzy probabilities.

16.2.6   Chapter 8

In this chapter we studied continuous fuzzy random variables. We looked at the following types of continuous fuzzy random variables: the fuzzy uniform, the fuzzy normal; and the fuzzy negative exponential. In each case we discussed how to compute fuzzy probabilities and how to determine their fuzzy mean and their fuzzy variance.

16.2.7   Chapter 9

Here we are interested in the "probabilistic" inventory control problems, in particular those with probabilistic demand. Suppose demand is modeled as a normal probability density. The mean and variance must be estimated from past data and are good candidates for fuzzy number values. We looked at two cases using the fuzzy normal density function for demand. The first case was a simple single item, one period, inventory model with fuzzy demand using the decision criterion of minimizing expect costs. The second case we expanded the model to multiple periods and now we wish to maximize expected profit.

16.2.8 Chapter 10

This chapter extends the results of Chapter 4 and 8 to multivariable fuzzy probability (mass) density functions. We studied fuzzy marginals, fuzzy conditionals and fuzzy correlation. Also, we discussed the fuzzy bivariate normal density.

16.2.9 Chapter 11

Here we had an application of a discrete fuzzy trinomial probability mass function, with its fuzzy marginals and its fuzzy conditional probability mass functions. The other application was using a joint discrete fuzzy probability distribution, a fuzzy Poisson and a fuzzy binomial, in reliability theory.

16.2.10 Chapter 12

Let X be a fuzzy random variable and $Y = f(X)$. In this chapter we show, through five examples, how to find the fuzzy probability (mass) density function for Y.

16.2.11 Chapter 13

This chapter generalizes Chapter 12. If X_1 and X_2 are fuzzy random variables and $Y_1 = f_1(X_1, X_2)$, $Y_2 = f_2(X_1, X_2)$, find the joint fuzzy probability (mass) density function for (Y_1, Y_2). We first look at how to solve the problem, through three examples, when the transformation is one-to-one. Then we see how to solve the problem, for two of the order statistics, when the transformation is not one-to-one.

16.2.12 Chapter 14

Limit laws are important in probability theory. We present only one in this chapter, the law of large numbers, using the fuzzy normal.

16.2.13 Chapter 15

We define the fuzzy moment generating function. We use this, just like the crisp moment generating function is used in crisp probability theory, to find the fuzzy probability (mass) density function for the sum of independent, identically distributed, fuzzy random variables.

16.3   Research Agenda

16.3.1   Chapter 3

What is needed is a numerical optimization method for computing the max/min of a non-linear function subject to both linear and inequality constraints in order to find the α-cuts of fuzzy probabilities. It would be nice if this numerical method could be coupled with a graphical procedure so we can graph these fuzzy probabilities and have the ability to export these graphs to $LaTeX2_\epsilon$. More work can be done on the basic properties of our fuzzy probability including fuzzy conditional probability and fuzzy independence.

16.3.2   Chapter 4

There are other discrete fuzzy random variables to be considered including: fuzzy uniform; fuzzy geometric; fuzzy negative binomial, ...

16.3.3   Chapter 5

There are lots of other queuing systems to investigate.

16.3.4   Chapter 6

There are many other results on Markov chains to study using fuzzy probabilities and restricted fuzzy arithmetic.

16.3.5   Chapter 7

There are other decision making under risk models that can be investigated using fuzzy probabilities and restricted fuzzy arithmetic.

16.3.6   Chapter 8

There are other continuous fuzzy random variables to study including: the fuzzy beta; the fuzzy Chi-square; the fuzzy gamma, ...

16.3.7   Chapter 9

There are lots of other probabilistic inventory control problems to consider using fuzzy probabilities.

16.3.8   Chapter 10

We only discussed the joint fuzzy probability density for two fuzzy random variables. Extend to $n \geq 3$ fuzzy random variables. Also, related to correlation is linear regression. Is fuzzy regression related to fuzzy correlation?

16.3.9 Chapter 11

There are other applications of fuzzy probabilities to reliability theory.

16.3.10 Chapter 12

Develop a general theory of finding the fuzzy probability (mass) density of $Y = f(X)$. What to do in Example 12.3.1 where we could not solve the problem?

16.3.11 Chapter 13

Generalize to $n \geq 3$ fuzzy random variables. Also derive the fuzzy t-distribution and the fuzzy F-distribution.

16.3.12 Chapter 14

Develop more general limit laws for fuzzy probability. Is there a fuzzy central limit theorem?

16.3.13 Chapter 15

Extend the fuzzy moment generating function to more than one fuzzy random variable. Develop the theory of limiting fuzzy moment generating functions.

16.4 Conclusions

Classical probability theory is the foundation of classical statistics. We propose fuzzy probability theory to be the foundation of a new fuzzy statistics.

Index

List of Figures

List of Tables

GPSR Compliance
The European Union's (EU) General Product Safety Regulation (GPSR) is a set
of rules that requires consumer products to be safe and our obligations to
ensure this.

If you have any concerns about our products, you can contact us on

ProductSafety@springernature.com

In case Publisher is established outside the EU, the EU authorized
representative is:

Springer Nature Customer Service Center GmbH
Europaplatz 3
69115 Heidelberg, Germany